A Science-Based Critique of Epistemological Naturalism in Quine's Tradition

Reto Gubelmann

A Science-Based Critique of Epistemological Naturalism in Quine's Tradition

palgrave
macmillan

Reto Gubelmann
University of Zurich
Zurich, Switzerland

ISBN 978-3-030-24523-8 ISBN 978-3-030-24524-5 (eBook)
https://doi.org/10.1007/978-3-030-24524-5

This Palgrave Macmillan imprint is published by the registered company Springer Nature Switzerland AG
The registered company address is: Gewerbestrasse 11, 6330 Cham, Switzerland

Für Nathalie

Preface

In this book, I venture a critical assessment of epistemological naturalism. Proponents of this position hold that there is no knowledge outside of science, since the empirical-scientific method (or methods) of justifying claims or theories is without alternative. Therefore, if epistemology is to deliver knowledge, it has to itself become immanent to science.

Often, critics of epistemological naturalism argue that the doctrine cannot accommodate some important aspect of human existence, say altruistic behavior. One weakness of this approach is that epistemological naturalists can always adopt an eliminativist or reductivist attitude with regard to the phenomenon in question: naturalists can argue that altruism either does not exist, or that it can be reduced to biological functions such as the reproductive success of groups.

This book is different. Rather than focusing on phenomena that are traditionally conceived to be external to empirical science, it focuses on empirical science itself. With regard to this phenomenon, epistemological naturalists cannot adopt an eliminativist position, and adopting a reductivist one comes at a considerable cost.

Furthermore, I am trying to meet my naturalistic counterpart on her own ground not only with regard to the subject matter (empirical science), but also with regard to the epistemic way in which I approach this subject matter: I try to support my central claims and arguments with detailed discussions of first-order scientific data. When arguing that it is an open question whether empirical science is able to explain how our access to empirical reality supports our scientific theorizing, I am making extensive use of first-order scientific results and experiments (such as habituation and preferential looking experiments). Even when my claim is squarely metaphilosophical, such as when I am urging that the scientific realism debate belongs to philosophy and not to science, I am making my case from within science, arguing that the debate transcends the *bona fide* epistemic practices of the sciences, and I am supporting my case with detailed analyses of actual discussions in the history of natural science.

When beginning the research for this study, one of my basic (but rather vague) working hypotheses was that the richness and diversity of actual scientific research militates against the austere and monolithic nature of epistemological naturalism. I think that my discussion of contemporary perceptual psychology in Sect. 4.2 and of Perrin's experiments to establish the reality of atoms in Chapter 9 support more specific versions of this general working hypothesis.

Hence, this book's contribution to the ongoing debate about epistemological naturalism—and about the metaphilosophy of epistemology in general—consists in arguing that there are grounds within science to question epistemological naturalism. By reviewing insights from linguistics, psychology as well as from the history of natural science, I argue that science itself points beyond itself to epistemic practices that are clearly distinct from empirical science.

Part of the research on which this book is based has been supported by The University of Zurich's Candoc grant number FK-13-068 as well as The Swiss National Science Foundation's doc.Mobility grant number P1ZHP1-161979.

Parts of Chapters 2, 3, and 7 draw on and elaborate ideas previously published in "From Shared Stimuli to Preestablished Harmony: The Development of Quine's Thinking on Intersubjectivity and Objective Validity", *HOPOS, 9*, DOI: https://doi.org/10.1086/703253. Figure 4.2 is reproduced with permission from Yang, Jiale, So Kanazawa,

Masami K. Yamaguchi, and Ichiro Kuriki (2013). "Investigation of color constancy in 4.5-month-old infants under a strict control of luminance contrast for individual participants". In: *Journal of Experimental Child Psychology* 115.1, p. 130.

I am grateful to the following people for their comments and criticism of various ideas and arguments in this book: Karim Bschir, Tyler Burge, Johannes Corrodi, David Dolby, Charles Djordjevic, Matthias Egg, Zacharias Flouris, Hans-Johann Glock, Gary Hatfield, Penelope Maddy, Christoph C. Pfisterer, Nicole Rathgeb, Katia Saporiti, Peter Schulthess, Kyle Stanford, Marco Toscano, Sander Verhaegh, Christian Weibel, Sebastian Wyss, and the audiences at the Free University of Amsterdam, the Universities of Helsinki, Pretoria, Vilnius, and Zurich.

St. Gallen, Switzerland Reto Gubelmann

Contents

About the Author

Reto Gubelmann is a part-time lecturer in Philosophy at the Universities of Zurich, Switzerland, and St. Gallen, Switzerland, and head of program development at the University of St. Gallen. He specializes in epistemology, metaphilosophy, philosophy of science, and Quine.

List of Figures

1
Introduction

Ever since the beginning of modern science and its emancipation from "natural philosophy", philosophers have marveled at the successes and the epistemic[1] discipline of modern science. Immanuel Kant, in his preface to the second edition of the "Critique of Pure Reason", states that, in contrast to natural science, philosophy has been unable to walk the secure path of a science. Instead, it is, and has always been, a mere groping among concepts (Kant 1998 [1987], pp. XIVf.). In response to this untenable situation, Kant intends to refashion philosophy so that it displays the desirable epistemic properties of natural science.

Kant's conviction that modern natural science poses a challenge to the self-understanding of philosophy is by no means an exception; rather, it is the rule. Given the lack of consensus, progress and discipline of philosophy,

[1] My distinction between epistemic and epistemological matters distinguishes the actual practice of justifying claims and attaining knowledge from reflecting and theorizing about this practice. Epistemic questions are questions regarding the justification of a given claim or position; they are internal to the game of giving and asking for reasons. For example, to maintain that the epistemic standing of the claim that god exists is weak is to maintain that the claim has little warrant, or justification. In contrast, epistemological questions are questions about this game; they are answered with reflections and theories about central concepts or principles of the game. By drawing this distinction, I do by no means dispute that many epistemic debates evolve into epistemological debates.

R. Gubelmann, *A Science-Based Critique of Epistemological Naturalism in Quine's Tradition*, https://doi.org/10.1007/978-3-030-24524-5_1

when compared to empirical science, many philosophers have begun to worry whether the supposed insights provided by philosophical inquiry are trustworthy. Indeed, it has become a question of serious consideration whether there is anything deserving of the title of knowledge outside of empirical science.

Epistemological naturalism, the topic of this book and one of the most influential epistemological positions in contemporary western philosophy, answers this question with a clear "no". The position stands in the tradition championed by Kant: epistemological naturalists seek to refashion epistemology to set it on the secure path of a science. What clearly distinguishes epistemological naturalism from proposals such as Kant's is that naturalism requires that epistemology adopt the same epistemic methods that are in use in the natural sciences. Following W. V. O. Quine (1981, p. 72), in a first approximation, epistemological naturalism can be characterized as rejecting any dualism in epistemology. In particular, it rejects the idea that epistemology and natural science have different epistemic resources and methods.

What is the general appeal of epistemological naturalism? Mostly, epistemological naturalists share the basic perspective of scholars like Kant: when compared with the epistemic merits of empirical science, philosophy, as traditionally practiced, seems to have little to recommend itself. Additionally, epistemological naturalists are typically particularly impressed by the instrumental and predictive success of empirical science. In the past four centuries, this success has been one of the most noteworthy phenomena in human intellectual life. From Newtonian physics to computational biology, from immunology to meteorology, empirical sciences have delivered theories that have vastly increased our power over our environment as well as our bodies, and they have delivered successful predictions of phenomena such as cosmic background radiation that were long thought to be far beyond the scope of human understanding.

As a consequence, it might seem natural to expect that this conquest of empirical science has no principled limits. It seems natural to expect that the epistemic methods of science (whatever they may be precisely), can be applied with equal success to any subject matter whatsoever.

The study presented in this book critically assesses the prospects of this naturalistic expectation; it does so in a notorious philosophical way, by focusing the attention on the subject itself: Can empirical science come to terms with its epistemic self, can it address it, using its tried and tested methods, as a phenomenon with (prospects of) the same success that it has had with so many other phenomena? With regard to these questions, that is, when the focus lies on science itself, the epistemological naturalist is committed to showing that empirical science is *epistemically self-sufficient*. The goal is to show that science does not need any other epistemic means than its own scientific ones—even in epistemological matters, which include examining and justifying its own way of justifying theories and hypotheses.

Note that the target of my critical analysis is not what is traditionally conceived as empirical science itself, but rather a certain epistemological position regarding science. According to this position, reflection about topics like scientific verification, truth, or possible conflicts between scientific and common-sense insights are to be addressed using essentially the same epistemic means that are used in everyday scientific inquiry.

In the literature, epistemological naturalism (see Horwich 2014, p. 38) is generally conceived as the position that "[o]nly the scientific method can deliver genuine knowledge", which of course echoes the claim phrased above, according to which there is no knowledge outside of science. This is not to say that there is only one version of epistemological naturalism. Rather, there are at least as many versions of epistemological naturalism as there are positions on how exactly science justifies its hypotheses and theories. It will soon become clear that Quine steadfastly defends an empiricist conception of how scientific justification works, and that he thinks he can support this conception in an austerely scientific way.

Epistemological naturalism is commonly distinguished from ontological and methodological naturalism, compare Glock (2003, pp. 27–28). What unites all of the three general kinds of naturalism is the conviction that philosophy should imitate science, although they differ from each other by imitating science in an epistemological, an ontological, or a methodological dimension.

Plantinga (2002, p. 1) conceives of naturalism as an ontological thesis, namely as the rejection of the existence of any supernatural beings, such as a god. Stricter versions of ontological naturalism, such as the one endorsed by Rosenberg (2014), amount to variants of materialism, mechanism, or

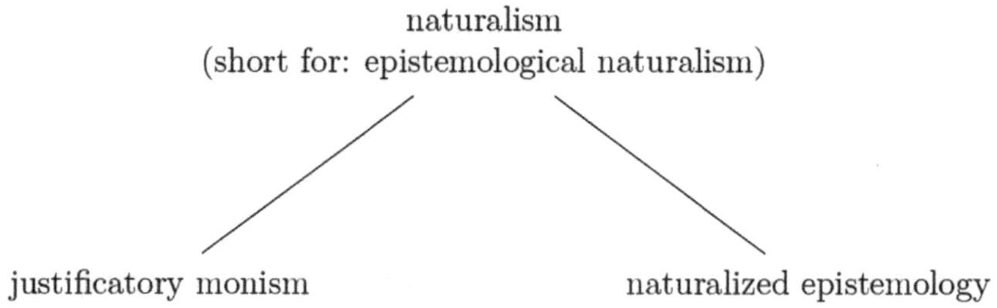

Fig. 1.1 An overview on the terminology used in this book (*Source* Author)

physicalism. I will show that Quine endorses a sophisticated version of physicalism, and hence a version of ontological naturalism.

Methodological naturalists urge that philosophy adopt the methods of science (compare Papineau 2015). However, these methods are almost always the methods of justifying claims, which means that, for the most part, methodological naturalism is epistemological naturalism by another name.[2]

In the remainder of this introductory chapter, I develop my conception of epistemological naturalism (Sect. 1.1), and I provide an overview on the content and the general argument of this book (Sect. 1.2).

1.1 Epistemological Naturalism

Epistemological naturalism is best conceived as consisting of two elements, one of them being a central claim, defining a central structural property of the position, what I call 'justificatory monism'. The other element constitutes a positive research agenda, namely the project of pursuing epistemology as a branch of empirical psychology, what I call 'naturalized epistemology'. Figure 1.1 gives an overview on this terminology.

In the following, I introduce justificatory monism as well as naturalized epistemology, and I detail their relationship. I do so with a focus on

[2]I take it that what Glock (2003, pp. 27–28) calls metaphilosophical naturalism largely coincides with methodological naturalism: he (ibid.) considers Quine's methodological monism as an instance of metaphilosophical naturalism.

Quine, the pioneering theoretician of the position, while also introducing the position of Penelope Maddy, whose subtle and innovative position will be discussed throughout the book.

1.1.1 Its Central Claim: Justificatory Monism

Naturalized epistemology is motivated by the conviction that epistemological reflection about natural science must be conducted using the same epistemic means that are used in first-order scientific inquiry. This conviction, in turn, is typically grounded in the belief that empirical-scientific justification is the only kind of justification available (this being a sufficient, but not a necessary condition for naturalized epistemology, see below, Sect. 1.1.3). I call this conviction, according to which there is only one basic way to justify any claim whatsoever, thus rejecting any kind of fundamental epistemic dualism, *justificatory monism.*

Note that the bare claim of justificatory monism does not specify what empirical-scientific justification amounts to—it does not even specify that the relevant kind of justification should be empirical. Hence, as such, the claim is almost hopelessly unspecific. What I call *versions of justificatory monism* involve a specific conception of justification. Furthermore, this book is about empiricist versions of justificatory monism, that is, versions of justificatory monism whose conception of scientific verification is recognizably empirical.

Note that, throughout the book, I am using "(empirical) verification" interchangeably with "(empirical) justification"—even though "empirical verification" is more common than "empirical justification". This is because I do not want, even on the linguistic level, to presuppose what is at issue. My naturalistic counterpart maintains that there is only one sort of justification; by systematically using, say, "justification" in the context of deductive arguments, and "verification" in the context of empirical support, I would implicate from the start that my counterpart is wrong.

Quine defines his justificatory monism in opposition to the view that there is an epistemically higher mode of inquiry, what Quine calls 'first philosophy':

> The fifth move, finally, brings naturalism: abandonment of the goal of a first philosophy. It sees natural science as an inquiry into reality, fallible and corrigible but not answerable to any supra-scientific tribunal, *and not in need of any justification beyond observation and the hypothetico-deductive method* [emphasis RG]. (Quine 1981, p. 72)

In this passage, Quine characterizes naturalism by means of two claims. Naturalism rejects the idea that first philosophy constitutes a tribunal to which science must answer. Considering the context, the tribunal probably does not judge the moral or legal standing of science, but rather its epistemic standing. Hence, first philosophy is in a position to decide whether the insights obtained by science are sufficiently warranted. That naturalism rejects this idea means that there is no higher court with the authority to judge the insights of science. In other words, the verdicts of science are final.

The second characteristic claim of naturalism is contained in the part of the passage that I have emphasized: naturalism rejects the idea that there is a way of justifying claims that is not reducible to observation and the hypothetico-deductive method, but still necessary for science. This means that naturalism rejects the claim that science needs a way of justifying claims that is strictly distinct from the one it, science, can provide itself.

The two claims are closely connected. If science was answerable to a supra-scientific tribunal, then it would need non-scientific means of justification to stand its ground before this tribunal. Conversely, if science would need non-scientific ways of justifying some of its claims, then it is hard to see how the discipline furnishing these ways would not thereby become a first philosophy to science—at least with regard to the relevant claims. Still, the first claim implies an epistemic hierarchy between first philosophy and science that is not directly present in the second. Hence, while the first claim clearly implies the second, it is not clear whether the second one implies the first.

The focus of this book lies on the second characteristic claim of naturalism, as Quine elaborates it in this passage. As my overall goal is to develop a critique of epistemological naturalism, it is more modest to focus on this second claim than to address the first one. To refute the second characteristic claim, it is sufficient to show that science is not epistemically self-sufficient, that is, to show that there are questions pertaining to science, including epistemological questions, that science cannot answer by itself. To refute the first characteristic claim would require showing that there is an epistemically higher mode of inquiry, which would obviously be a more ambitious goal. Hence, what I call justificatory monism (and what Quine calls naturalism[3]) *requires that any statement whatsoever that is necessary for science can be justified by scientific justification.*

Of course, much depends in this definition on the precise meaning of 'necessary for science' and 'scientific justification'. I discuss these two expressions in their order of appearance. First, while, broadly speaking, many statements are necessary for science, I focus on statements that are necessary for science insofar as they specify standards for empirical justification. These standards define the game of science (this is Quine's metaphor, see below, Sect. 6.2.2) and are therefore indispensable for the practice of scientific inquiry. According to justificatory monism, these standards are justified in the same way that they themselves determine.[4]

Second, according to Quine, 'scientific justification' is to be understood in a strictly empiricist-holistic way. Whether a given statement is justified or not depends entirely on whether it contributes to the successful prediction of empirical data (compare Quine 1992, p. 20). Thus, Quine's version of justificatory monism, which I call his *empiricist justificatory monism*, can be specified as follows: any statement whatsoever that is necessary for science can be justified by its contribution to the successful prediction of empirical data.

The basic thrust of Maddy's naturalism agrees nicely with Quine's rejection of first philosophy, that is, with the rejection of any strict distinction between science and philosophy as well as of the idea that science is in

---

[3]I use the term 'justificatory monism' instead of Quine's term 'naturalism' for two reasons. First, 'naturalism' is—mostly due to Quine's influence—currently used in so many different ways that it has lost any distinctive meaning. Second, I hope that the term 'justificatory monism' serves to highlight the aspects of Quine's naturalism that I am focusing on.

[4]Hence, I disagree with Keil (2003, p. 277), who maintains that "it was far from clear what he [Quine, RG] did ban under the name of prior [i.e. first, RG] philosophy", and also that "[t]here are not many philosophers left at whom Quine's criticism of philosophical apriorism could be aimed." Quine's justificatory monism excludes any position according to which reflection on the correct explication of empirical-scientific justification is not itself bound by this conception. Such positions are still held by a significant part of contemporary philosophers of science.

need of, or answerable to, any non-scientific mode of reflection and justification. Hence the term that Maddy uses for her idealized naturalistic inquirer, namely the "Second Philosopher". Apart from this fundamental agreement, however, Maddy's version of justificatory monism, called "Second Philosophy", differs in interesting ways from Quine's.

In explicit terms, Maddy explains Second Philosophy merely negatively. Second Philosophy is fundamentally opposed to any epistemic two-level conceptions of the relationship between philosophy (and epistemology in particular) and science. According to Maddy (2007, pp. 56–64), the archetype of such a two-level view is Kant's transcendental philosophy: on her view, Kant's distinction between the realm of empirical science and the domain of transcendental inquiry implies that the former has nothing whatsoever to contribute to the latter. In positive terms, however, Maddy refuses to specify her justificatory monism, that is, to give a positive, explicit epistemological specification of what scientific justification consists in. Maddy refuses to do so because she thinks that it is probably an impossible task (Maddy 2007, p. 1) and because she, ultimately, holds that it is incompatible with naturalism to give a hard and fast specification of what empirical justification consists in (see esp. Maddy 2007, pp. 91f.). Hence, Maddy's justificatory monism rejects any conception of philosophy that is epistemically strictly distinct from the realm of science—without specifying positively what is epistemologically distinctive about this realm.

While Maddy does not have an explicit, clear-cut specification of what empirical justification, and hence empiricist justificatory monism, consists in, she embraces empiricist justificatory monism in the practice of her naturalism. In particular, she uses the notion of the empirical to make sense of what a two-level view amounts to (see, e.g., Maddy, 2007, p. 4). Furthermore, she emphasizes that her epistemology, in contrast to Kant's, enjoys solid empirical support (Maddy, 2007, p. 6). Finally, as I will argue in Chapter 9, Maddy uses a, perhaps problematic, distinction between empirical and non-empirical aspects of the atoms debate to position herself vis-à-vis this debate.[5]

[5]Further passages where Maddy implicitly relies on empiricist justificatory monism are: Maddy (2007, p. 262), where she makes clear that it is not conceivable that any other evidence than empirical evidence could play a role in developmental psychology; Maddy (2007, p. 279), where

Leaving more specific questions aside, it is not problematic from the start to refuse to explicitly state a certain conceptual distinction while one's intellectual practice displays this same distinction at work. Furthermore, Maddy (2007, p. 1) is fully aware of this aspect of her position. She admits that she does not have a general definition, of the kind of Quine's empiricism, for the realm of science; what she does instead is introduce the Second Philosopher as a character and then "proceed by describing her thoughts and practices in a range of contexts". Hence, while her explicit characterization of her position is merely negative—consisting in a rejection of any two-level views à la Kant—it seems apt to conceive her position as an implicit and unspecific kind of empiricist justificatory monism—all the while keeping in mind that the Second Philosopher would never assume the meta-position to make this assertion.

1.1.2 Its Practice: Naturalized Epistemology as Empirical Psychology

Naturalized epistemology is the only remaining way to practice epistemology, once one has accepted justificatory monism. It is the attempt to pursue epistemology as a part of empirical science, namely as empirical psychology; to understand, using solely the epistemic practices of the empirical sciences, the central and basic cognitive abilities necessary for empirical science, with the ultimate goal to determine the kind of evidential support that science can enjoy.

This means that naturalized epistemology is oriented towards two main issues: (1) what is our most direct and unmediated cognitive contact with empirical reality, and what are the cognitive abilities of humans that are both of central importance for scientific inquiry and that are in most direct

she acknowledges, in striking analogy to Quine, that even the laws of logic can be revised in light of empirical evidence; Maddy (2007, p. 410), where she agrees with Mark Lange that interpretations of physical theories that allow for new empirical predictions thereby qualify as scientific theories, instead of being philosophical speculation; Finally, Maddy (2007, p. 88) distinguishes her position from Carnap's by rejecting the very idea that there is any realm of inquiry beyond the empirical one: "what she [the Second Philosopher, RG] wants, of course, is a defense based on empirical study of human language use. In contrast, Carnap explicitly distinguishes his pursuit of 'the logic of science' from the empirical study of language".

and unmediated cognitive contact with empirical reality? (2) How, and to what extent, does this contact support our scientific theorizing?

The basic idea of naturalized epistemology is by no means new: for millennia, researchers have studied the way in which we "take in information" about empirical reality, how our mind processes this information so that it can be used as evidence for our beliefs and, ultimately, for our scientific theories. What distinguishes naturalized epistemology from these traditional approaches is that it explicitly rejects most traditional modes of inquiry—typically versions of a priori reflection—for the purposes of this project.

To illustrate this very general description, I now sketch the versions of naturalized epistemology developed by W. V. O. Quine and Penelope Maddy. Additionally, I briefly introduce research by Tyler Burge, who, as I will show, occupies an interesting middle ground between Quine and Maddy—without subscribing to naturalized epistemology.

In his seminal article "Epistemology Naturalized", Quine introduces the aim of naturalized epistemology as finding out how "evidence relates to theory, and in what ways one's theory of nature transcends any available evidence" (Quine 1969, p. 83). Accordingly, naturalized epistemology examines the relationship between evidence and theory, and the principled limits that follow from this examination for the extent to which a scientific theory can be warranted by evidence.

According to Quine, the only empirical data or evidence available to us is given by patterns of sensory stimulations of our nerve endings. He considers this itself a scientific insight, and one that enjoys a particularly high degree of justification (for details and references, compare Chapter 2 below, in particular Sect. 2.2.1). Such patterns of sensory stimulations, in turn, can only be directly associated with *holophrastic* observation statements, that is, with statements taken as one-word sentences that are not ontologically committed or (proto-)logically structured. This is part of Quine's answer to the issues raised in (1) above: our most basic, unmediated contact with empirical reality is given by patterns of sensory stimulation of our nerve endings. Hence, our entire scientific edifice can stand on no more substantial empirical foundations than these patterns of sensory stimulations.

Quine's answer to (2), that is, his account of how this contact with empirical reality does in fact support our scientific theorizing, proceeds by examining the ontogenesis of language: he intends to elucidate the connection between theory and evidence by studying how a typical human child develops a primitive scientific theory based on this empirical contact with the world. This presupposes two basic convictions of Quine's overall philosophy. First, Quine holds that every language constitutes a primitive scientific theory of the world. Second, he maintains that, when a scientist is testing her theories and when an infant is learning his first language, both the scientist and the infant have the same basic kind of evidence available. Epistemologically speaking, the scientist and the infant are sitting in the same boat.

This means that Quine can study the genesis of science, and hence its justification, by studying how a human infant acquires his first language.

In the following passage, Penelope Maddy describes the epistemological research of her idealized naturalistic inquirer, the Second Philosopher:

> In general, the Second Philosopher's epistemological investigations take the form of asking how human beings – as described in biology, physiology, psychology, linguistics, and so on – come to have reliable beliefs about the world – as described in physics, chemistry, botany, astronomy, and so on. (Maddy 2014, p. 100)

Here, Maddy sketches the basically Quinean project of explaining scientifically how humans could have developed a scientific worldview. Hence, Maddy clearly shares the basic goal of Quine's naturalized epistemology.

Maddy also follows Quine in making developmental psychology the central discipline for naturalized epistemology. The basic idea behind this is that pre-linguistic infants have a kind of access to empirical reality that is not mediated by language, culture, or scientific preconceptions and hence much more basic and direct than the kind of access that adults have.

However, as I will detail below (Chapter 5), her answers to (1) and therefore also to (2) are about as different from Quine's as one can imagine. Maddy argues, based on experiments pioneered by developmental psychologist Elisabeth Spelke, that pre-linguistic human infants possess concepts and are able to judge the possibility or impossibility of certain simple physical events. This means that, rather than mere patterns of sensory stimulation, Maddy claims that our most basic contact with empirical reality is already conceptually structured; furthermore, she holds that infants already reason according to a simple kind of logic to judge the possibility or impossibility of such simple physical events. In answer to (2), Maddy's position implies that scientific theorizing ultimately rests on the same basic kind of contact with the world that pre-linguistic infants already possess, namely conceptually structured perceptions that are embedded in logical thought.

Tyler Burge, finally, occupies a middle position between Quine and Maddy when it comes to his answer to (1) (see below, Chapter 4). He argues that the most basic cognitive contact with the world that humans and many other animal species have is given through perceptual representations. While these representations are not conceptually structured (a claim that Maddy ventures with regard to the thoughts of pre-linguistic infants), they display an amount of structure and sophistication that qualitatively distinguishes them from Quine's patterns of sensory stimulations. It will turn out that Burge's answer to (2) also occupies a middle ground between Quine's and Maddy's positions.

Sometimes during my study of naturalized epistemology, it will be useful to have a more concrete act of cognition in mind than the slightly amorphous notion of scientific theorizing. For this purpose, I will invoke human judgments. Corresponding to established usage (compare Glock 2010, pp. 25–26), I am understanding judgments as mental acts, resulting in mental entities that are mental analogues to states of affairs on the one hand and linguistic sentences on the other. These sentences involve predication, that is, the application of a general, repeatable property to some individual, concrete particular. With regard to this, naturalized epistemology faces the task, typically based on an explanation of the ontogenesis of judgment and hence of predication, to elucidate the evidential basis of these judgments (compare (2) above).

1.1.3 Some Structural Aspects

When practiced, epistemological naturalism becomes naturalized epistemology, a branch of empirical psychology and hence of empirical science. It may use findings of developmental psychology and linguistics, of cognitive and perceptual psychology and of any other empirical science to empirically elucidate the kind of evidence that human beings have available to theorize about the world. This evinces that naturalized epistemology, Quinean and other, is *inherently circular*: in its explanation of the emergence of science and of its evidential base, it employs the very scientific insights whose emergence it intends to investigate. Quine rightly argues that this circle is not vicious for him, as he does not intend to place science on a firmer foundation than science itself. Rather, his goal is to explain in a scientific way—and hence by presupposing and employing any scientific theories that he finds useful—how humans can have developed successful scientific theories, based on mere patterns of sensory stimulations as sources of information about the world.

This containment of naturalized epistemology within science results in reconceiving epistemology as a part of psychology, and it yields one direction of the much-debated "reciprocal containment" of natural science and epistemology (Quine 1969, p. 83): Epistemology is contained in science insofar as the former is a species of the latter. This containment of naturalized epistemology in science brings both rights and duties for naturalized epistemology: it can now make free use of all scientific insights, but it is bound by the epistemic standards of science (Quine 1974, p. 34).

The other direction of the reciprocal containment is given by the fact that naturalized epistemology investigates the relationship between scientific theory and evidence. As a consequence, the results of naturalized epistemology, even if it is only a part of empirical science, apply to empirical science as a whole: Should naturalized epistemology find, say, that humans have no empirical access whatsoever to empirical reality, this would have important and far-reaching consequences for empirical science as a whole.

After discussing the relationship between naturalized epistemology and empirical science, I now turn to the relationship between justificatory monism and naturalized epistemology. In any consistent naturalistic position, the two doctrines should be *mutually supportive*. Consider the architecture of Quine's epistemological naturalism. The question is: how do you scientifically defend a specific view of empirical-scientific justification? Quine's very plausible idea to accomplish this is: by showing that the respective view follows from the most solid scientific insights that are relevant to the issue. In other words, his naturalized epistemology is intended to scientifically support his empiricist version of justificatory monism. The same holds in the case of Maddy's justificatory monism: if her naturalized epistemology establishes that there are two fundamentally distinct ways to justify claims and to gain knowledge, one corresponding to traditionally philosophical, a priori reflection, the other corresponding to empirical research, then her justificatory monism would be seriously questioned.

Conversely, justificatory monism provides a sufficient—but not a necessary—condition to naturalize epistemology: If it is the case that the only basic way to justify any claim whatsoever is through empirical-scientific evidence, then the only possible way to pursue epistemology is in the form of naturalized epistemology, by relying on empirical-scientific evidence. It does not provide a necessary condition because there could be other reasons for a researcher to pursue epistemology in a naturalized manner. For instance, she could agree that metaphysics, when properly practiced, furnishes synthetic insights that are a priori warranted—but that these insights cannot apply on matters epistemological.

Note a consequence of the reciprocal support of justificatory monism and naturalized epistemology: specific versions of justificatory monism support specific versions of naturalized epistemology by determining what counts as empirically justified, and vice versa. For example, if one's version of justificatory monism is strongly focused on predictive success, this determination might exclude from the practice of naturalized epistemology these parts of evolutionary biology that are historical rather than predictive. The outcome of naturalized epistemology, in turn, by its very goal, determines the kind of evidence available to humans and thereby scientifically determines the right version of justificatory monism.

According to the schematic representation in Fig. 1.2, which elaborates on Fig. 1.1, justificatory monism and naturalized epistemology support each other, and both are located in the field demarcated by epistemological naturalism.

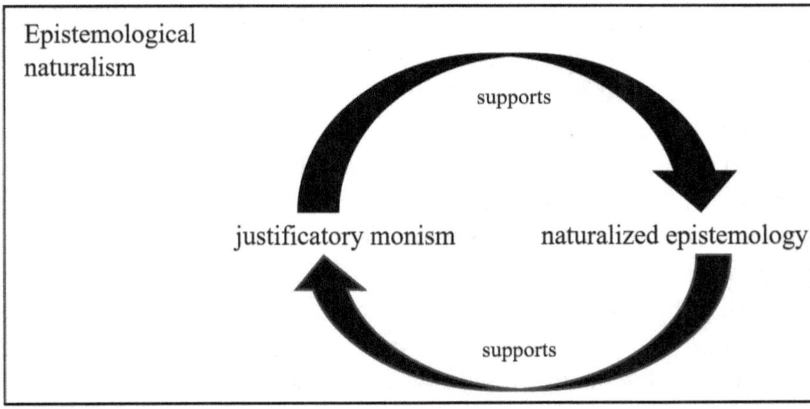

Fig. 1.2 The relationship of reciprocal support between justificatory monism and naturalized epistemology (*Source* Author)

To delineate the dialectical ground for a critical assessment of epistemological naturalism, in the remainder of this section, I detail the consequences of the failure of one of the two relata for the other.

First, assume that it turns out that naturalized epistemology is impossible in principle. What does this imply for epistemological naturalism and for justificatory monism? I suggest that the implications are threefold. First, it simply means that epistemological naturalism has no practice. The claim that the only way to justify any claim whatsoever is through empirical-scientific evidence together with the insight that empirical-scientific evidence is useless in matters epistemological implies that there cannot be a positive research project called epistemology. Second, it also implies that justificatory monism remains an empty claim, as it were: for an empiricist justificatory monist, the specific conception of empirical-scientific justification must be warranted by empirical science. If it turns out that empirical science cannot contribute anything in these matters, then this means that the justificatory monist cannot specify what empirical-scientific justification consists in. This ultimately means that she cannot say what distinguishes science from non-science. Third, the principled failure of naturalized epistemology would also imply that justificatory monism is not sustainable because it would mean that it cannot be justified: being a paradigm case of an epistemological claim, it would have to be abandoned if it would turn out that empirical-scientific justification cannot ground any epistemological claim (remember that empiricist justificatory monism claims that there is no justification beyond empirical-scientific justification).

Second, consider the case where naturalized epistemology is spectacularly successful, but it turns out that justificatory monism is untenable, say, because it turns out that it is possible to justify synthetic judgments purely a priori, by following the correct method of metaphysical theorizing. It appears that this can at most question the epistemic hegemony, as it were, of naturalized epistemology. In this case, it might be that epistemological claims can also be warranted non-empirically. This would mean that naturalized epistemology is no more the only way to achieve insight into epistemological matters, but it would not as such invalidate any insight arrived at by naturalized epistemology.

1.2 The Content and General Argument of the Book

The book is divided in two parts. In the first part, I examine the prospects of naturalized epistemology. The second part is dedicated to a critical assessment of empiricist justificatory monism.

Part I This first part consists of four Chapters, 2, 3, 4 and 5. Chapter 2 serves three purposes. I first detail the doctrines that form the background of Quine's naturalized epistemology, namely, on the one hand, what I call his eschatological physicalism, which compels him to adopt a kind of behaviorism in psychology and to ban any final causes, or purposes, from his account and, on the other hand, what I call Quine's source- and checkpoint-empiricism. Second, I sketch the elegantly physicalistic account that Quine gives in *Word and Object* of the infant's first steps into language. Third, I detail a problem that led him to abandon this account.

In Chapter 3, I reconstruct Quine's struggle with this problem, I delineate his final proposal to solve it, and I assess the extent to which this final solution still agrees with the background doctrines mentioned above. Then, I develop an internal critique of two central aspects of Quine's mature naturalized epistemology.

Chapter 4 is dedicated to Tyler Burge's account of the origins of objectivity. According to him, perceptual representation has a proto-predicational structure, is achieved even by non-mammals and studied by a *bona fide* scientific discipline. I discuss Burge's proposal with an eye looking backward and an eye looking forward, as it were. First, the account adds to the critique of Quine's own mature proposal by suggesting that Quine has been superseded by the progress of science. Second, I consider the merits of Burge's account as an attempt at naturalized epistemology in its own rights (regardless of whether Burge explicitly intends to contribute to this research project).

Chapter 5 discusses Penelope Maddy's own naturalized epistemology and thereby again questions Quine's mature proposal. Maddy's proposal assumes that pre-linguistic infants refer to objects and attribute properties to them (a view that I call *nativism*). This view is diametrically opposed to Quine's source-empiricism. I focus on the experimental paradigm on which Maddy relies to establish her nativism. This paradigm is called habituation and preferential looking, and it was pioneered by the developmental psychologist and outspoken nativist Elizabeth Spelke. At the end of Chapter 5, I sum up my insights regarding the prospects of naturalized epistemology.

Part II This part of the book is constituted by Chapters 6, 7, 8, and 9. In Chapter 6, I show how Quine's empiricist justificatory monism is based on his verification holism, which in turn presupposes his holophrastic conception of empirical data. I then point out that the problems outlined in my discussion of Quine's naturalized epistemology threaten to undermine Quine's empiricist justificatory monism. Finally, I argue that Quine's explication of empirical-scientific justification itself cannot be justified in the sense that it itself specifies. Given justificatory monism, this means that the explication is unjustified.

In Chapter 7, I argue that this yields a general problem for empiricist justificatory monism, one that can only be avoided by adopting scientific realism. I then introduce contemporary scientific realism, and I detail Quine's rejection of it, showing that it contradicts fundamental tenets of his empiricist justificatory monism.

In Chapters 8 and 9, I defend the claim that, as a matter of fact, scientific realism is unavailable to the empiricist justificatory monist, even if we abstract from the specifics of Quine's empiricist justificatory monism. The basic reason for this is that scientific realism, as a doctrine, is not something that can be refuted or defended by empirical-scientific evidence. I defend this claim by analyzing the so-called no-miracle arguments for scientific realism proposed by Stathis Psillos and Howard Sankey (Chapter 8) as well as Maddy's piecemeal defense of scientific realism (Chapter 9), which focuses on a specific point in the history of science, namely on Jean Perrin's famous experiments around 1910. Maddy argues that these experiments, together with theoretical work of Albert Einstein, directly and scientifically establish the reality of atoms.

In the concluding Chapter 10, I summarize the argument developed in the book, and I draw consequences for the prospects of justificatory monism and naturalized epistemology.

The book focuses on the versions of naturalized epistemology and justificatory monism developed by Quine and Maddy; given his prominence in current debates, some readers might miss the work of Alex Rosenberg. After all, he is one of the most prominent defenders of what he himself calls "scientism" (Rosenberg 1996, p. 4), a doctrine very similar to epistemological naturalism. There are three reasons for not including him within the critical focus of this book. First, his acceptance of a strict materialism (compare again Rosenberg 2014) as well as his emphasis on prediction as the ultimate touchstone for scientific theories (Rosenberg 1996, esp. pp. 22ff.) and his fundamental opposition to intentionality (Rosenberg 1999, esp. pp. 347ff.) strongly suggest that my critique of Quine's naturalized epistemology also applies to Rosenberg's. In contrast, Maddy's non-reductive, more permissive position seems much more promising. Furthermore, while having published extensively on the general philosophy of science as well as on the philosophy of biology, naturalized epistemology does not figure as centrally in his oeuvre as it does in Quine's or Maddy's.

Finally, I submit that, while Rosenberg's work has got the critical attention that it deserves, researchers have still failed to fully appreciate the subtlety and strength of Maddy's naturalism. This book hopes to contribute to changing this by putting Maddy's work, together with Quine's, in the center of critical attention.

References

Føllesdal, Dagfinn. 1973. Indeterminacy of Translation and Under-Determination of the Theory of Nature. *Dialectica* 27: 289–301.

Glock, Hans-Johann. 2003. *Quine and Davidson on Language, Thought and Reality*. Cambridge: Cambridge University Press.

———. 2010. Can Animals Judge? *Dialectica* 64: 11–33.

Horwich, Paul. 2014. Naturalism and the Linguistic Turn. In *Contemporary Philosophical Naturalism and Its Implications*, ed. Bana Bashour and Hans D. Muller, 37–43. New York: Routledge.

Kant, Immanuel. 1998 [1987]. *Kritik der Reinen Vernunft*. Hamburg: Meiner.

Keil, Geert. 2003. "Science Itself Teaches". A Fresh Look at Quine's Naturalistic Metaphilosophy. *Grazer Philosophische Studien* 66: 253–280.

Maddy, Penelope. 2007. *Second Philosophy. A Naturalistic Method*. Oxford University Press.

———. 2014. A Second Philosophy of Logic. In *The Metaphysics of Logic*, ed. Penelope Rush. Cambridge: Cambridge University Press, pp. 93–109.

Papineau, David. 2015. Naturalism. In *The Stanford Encyclopedia of Philosophy*, ed. Edward N. Zalta. Stanford: Stanford University. https://plato.stanford.edu/archives/win2016/entries/naturalism/. Accessed 5 August 2017.

Plantinga, Alvin. 2002. The Evolutionary Argument Against Naturalism. In *Naturalism Defeated? Essays on Plantinga's Evolutionary Argument Against Naturalism*, ed. James Beilby, 1–12. Ithaca: Cornell University Press.

Quine, Willard Van Orman. 1969. Epistemology Naturalized. In *Ontological Relativity and Other Essays*, 69–90. New York: Columbia University Press.

———. 1974. *The Roots of Reference*. La Salle: Open Court Publishing.

———. 1981. Five Milestones of Empiricism. In *Theories and Things*, 67–72. Cambridge: Harvard University Press.

———. 1992. *Pursuit of Truth*, 2nd ed. Cambridge: Harvard University Press.

Rosenberg, Alex. 1996. A Field Guide to Recent Species of Naturalism. *The British Journal for the Philosophy of Science* 47: 1–29.

———. 1999. Naturalistic Epistemology for Eliminative Materialists. *Philosophy and Phenomenological Research* 59: 335–358.

———. 2014. Disenchanted Naturalism. In *Contemporary Philosophical Naturalism and Its Implications*, 17–36. New York: Routledge.

2

Quine I: The Bold Physicalism of *Word and Object*

This present chapter as well as the next are dedicated to Quine's naturalized epistemology, which is the attempt to scientifically elucidate both the justification and the genesis of science and thus to confirm Quine's holophrastic conception of empirical data.

I focus on a small part of Quine's overall account, namely on how the infant acquires her first observation sentences. The reason for this focus is twofold. First, Quine (1974, pp. 35–36) declares that the infant's first steps into language are well-suited for behavioristic studies. In contrast, he hedges his account of the later stages of language learning, pointing to the difficulties of finding suitable triggers for specific verbal behavior.[1] Hence, a critique of Quine's naturalized epistemology that aims at questioning claims to which Quine is fully committed should be directed at the first stages of language learning, and not at the later ones.

The second reason for focusing on this first stage of first-language acquisition is systematic: it is by studying observation sentences, and their

[1]Compare the following passage: "Infant learning is a bright domain, and there behavioristic psychology blooms. The beginnings of language are learned ostensively. The needed stimuli are right out there in front, and mystery is at minimum. The old-time talk of ideas, ideas grasped and ideas conveyed, is given out without a wrench. Subtleties and obscurities crowd in, however, when we press to less primitive stages of language learning" (Quine 1974, p. 35).

R. Gubelmann, *A Science-Based Critique of Epistemological Naturalism in Quine's Tradition*, https://doi.org/10.1007/978-3-030-24524-5_2

relationship to patterns of sensory stimulation, that Quine thinks one can really study the extent of justification for science. Here is how Quine expresses this aspect of his epistemology in *The Roots of Reference* and then in *Pursuit of Truth* (1992):

> Observation sentences are the gateway to language, as to science. (Quine 1974, p. 40)

> Observation sentences are the link between language, scientific or not, and the real world that language is all about. (Quine 1992, p. 5)

To examine the soundness of Quine's account of how observation sentences can serve both as the gateway to language and as checkpoints for science, this chapter is structured as follows. First, I delineate the basic doctrines and convictions of Quine that form the background of his explanation of how infants acquire their first observation sentences. These doctrines and convictions are a sophisticated version of physicalism, which forbids Quine to use final causes in naturalized epistemology and naturally suggests behaviorism as psychological methodology, as well as his source-empiricism (Sect. 2.1). Note that these doctrines and convictions are admitted for naturalized epistemology because they are supposed to enjoy a high degree of justification, as it is specified in Quine's empiricist justificatory monism.

Then, I reconstruct Quine's account of the infant's acquisition of her first observation sentences, as he gives it in *Word and Object*, and I also show that the account—albeit elegant, boldly physicalistic, and highly innovative—suffers from a fatal flaw that Quine himself acknowledged a couple of years after the publication of *Word and Object*: what he proposed as the meaning of observation sentences, stimulus meaning, cannot provide the basis for communication (Sect. 2.2).

Before starting this examination, it is useful to distinguish Quine's (later) philosophy from *Word and Object* (1960) onward from his earlier views. *Word and Object* marks a watershed in several respects. Two of them are particularly important for present purposes. First, in *Word and Object*, Quine develops his holophrastic conception of empirical data to support the verification holism that he introduced in *Two Dogmas of Empiricism* (1951, in what follows often abbreviated by *Two Dogmas*).

There, he defended verification holism only negatively, by pointing out that all attempts at deducing specific theoretical claims from observation statements had failed.[2]

The second respect in which *Word and Object* marks a watershed is with regard to Quine's physicalism. Whereas in *On What There Is*, Quine (1980 [1948], p. 17)[3] proposed to develop two different conceptual schemes, a physicalistic and a phenomenalistic one, in *Word and Object*, he is clearly exclusively committed to the former. From *Word and Object* onward, these two doctrines—a holophrastic conception of empirical data and a version of physicalism—constitute basic pillars of his entire philosophical system throughout his published work. In this study, I examine Quine's philosophy in the form it takes from *Word and Object* onward—with only a few glimpses at his views before *Word and Object*.

2.1 The Theoretical Background of Quine's Naturalized Epistemology

In this section, I discuss the theoretical commitments that form the background of Quine's naturalized epistemology. First, I detail his subtle kind of physicalism (Sect. 2.1.1). Then, I introduce his rejection of final causation (Sect. 2.1.2) and his behaviorism (Sect. 2.1.3). Fourthly, I discuss his empiricism, which assigns two different functions to his holophrastic conception of empirical data: these data serve as our only source of information about the world and as ultimate checkpoints for scientific theorizing (Sect. 2.1.4).

Before focusing on each of these four background doctrines of Quine's naturalized epistemology in turn, I give an idea of how they work together in his naturalized epistemology. Consider how Quine describes the basic

[2] *On the Very Idea of a Third Dogma* contains an insightful passage, where Quine reflects on this development (Quine 1981, p. 40).

[3] A note on my way of quoting works by Quine: where the publication year of a given work by Quine that I am using differs from the year in which it was first published, I add the year of its first publication in square brackets. For instance, I am quoting Quine's *On What There Is* from the second edition of the collection *From a Logical Point of View*, which appeared in 1980. *On What There Is* was first published in 1948, hence I am referencing it as 'Quine (1980 [1948])'.

set-up of naturalized epistemology in this passage from *Epistemology Naturalized* (in what follows referred to as 'the EN passage'):

> Epistemology, or something like it, simply falls into place as a chapter of psychology and hence of natural science. It studies a natural phenomenon, viz., a physical human subject. This human subject is accorded a certain experimentally controlled input – certain patterns of irradiation of assorted frequencies, for instance – and in the fullness of time the subject delivers as output a description of the three-dimensional external world and its history. The relation between the meager input and the torrential output is a relation that we are prompted to study for somewhat the same reasons that always prompted epistemology; namely, in order to see how evidence relates to theory, and in what way one's theory of nature transcends any available evidence. (Quine 1969a, pp. 82–83)

In this passage, Quine sketches a rather bold research program. First, he clarifies the disciplinary status and the subject matter of this mode of inquiry: epistemology naturalized is a part of psychology, which is in turn part of natural science; its subject matter is the process through which a *physical* human subject develops a theory of the world from patterns of causal sensory stimulation—what Quine calls *input*. What is not explicit in the EN passage is that Quine's naturalized epistemology focuses on the emergence of reference to individual objects as a part of said description of the three-dimensional world (see Quine 1974, p. 123).

Quine's physicalism leads Quine to conceive of the object of his studies as a *physical* human subject. Quine's physicalism is also manifest in how the EN passage conceives of the input given to this subject: it does not consist in sense data or intentional representations, but rather in "patterns of irradiation of assorted frequencies".

Quine's behaviorism shows itself in the basic conception of the experiment: the passage sketches scientific theory construction as a very complicated stimulus-response pattern.

Quine's empiricism, finally, which is heavily influenced by his physicalism, allows him to describe the process as a rather simple input-output phenomenon: there is no room here to assume that not only the experimentally controlled, physicalistic input, but perhaps also some innate cognitive structures, or some substantial cognitive processing of the input, could be a fundamental source for the theory-construction of the physical subject.

Note, finally, two central commitments of this set-up for naturalized epistemology. First, for Quine, every language constitutes a primitive theory of the world, in particular, it contains a specific ontology. Hence, answering the question how humans acquire their first language also answers how they develop their first primitive theory of the world.[4] Second, note how the passage moves from genetic studies to epistemological insights. In the first part of the EN passage, the relationship between theory and evidence is studied for genetic insight: it focuses on the process by which the physical human subject develops the theory of the world based on nothing else than the sensory evidence. Then, Quine makes clear that the same process can also be studied to elucidate the justification available for scientific theories: the infant learns her first language based on the same evidence that the scientist uses to test her hypotheses. Therefore, the genetic study sketched in the EN passage promises insight into the means that are available to the scientist to test her theories, and the extent to which her theories are beyond such testing.

2.1.1 Eschatological Physicalism

In this section, I detail Quine's physicalism. First and foremost, Quine's physicalism marks a commitment to a certain discipline, namely physics. Quine's physicalism decrees that for disciplines other than physics it is never legitimate to theorize in a way that transcends the scope of state-of-the-art physical theories on an ontological and on an explanatory level.[5]

Physics Is Ontologically Basic. Here is how Quine spells out the ontological level of his physicalism in *From Stimulus to Science*:

[4]Here is how Quine moves from the former to the latter in *Roots of Reference*: "We have been speculating on the mechanisms of language learning, with emphasis on cognitive language. Our general objective was a better understanding of how scientific theory can have been achieved" (Quine 1974, p. 81).

[5]My discussion of Quine's physicalism is indebted to Hylton (2007, pp. 310–317). In particular, I agree with his clarification that physicalism, for Quine, does not exclude the existence of abstract objects such as sets. My overall evaluation of Quine's physicalism, in contrast, differs substantially from Hylton's.

> Physicalism, on the other hand, is materialism, bluntly monistic except for the abstract objects of mathematics. (Quine 1995, p. 15)

According to this passage, Quine's physicalism is a combination of a monistic materialism with the abstract entities needed for mathematics. This explains why Quine refused to posit intensional entities[6] in his naturalized epistemology.

There is another explanation of why Quine rejects intensional entities: they lack a clear criterion of identity (compare Gibson 1988, p. 13 for a canonical statement of this explanation). Quine uses his famous slogan "No entity without identity" in *Speaking of Objects* (Quine 1969 [1958], pp. 21–25) to exclude propositions and attributes from the conceptual scheme used for "official scientific business" (Quine 1969 [1958], p. 24). One could think that the same slogan is Quine's motivation against the mentalistic entities posited by cognitive science.

This alternative explanation may be exegetically tenable until the 1970s. However, from the 70s onward, at the latest, the question of identity criteria for mentalistic meanings is clearly not the main problem anymore, but rather Quine's physicalism. This reading is confirmed by Quine's struggle with the ontological status of electrons, and by his eventual acceptance of them into his ontology. Electrons lack determinate identity over time.

Quine did not accept electrons in his ontology without, as it were, putting up a fight, as the following passage from *Whither Physical Objects* shows. Immediately before the passage, Quine proposes to abandon the notion of a physical object altogether in favor of an ontology of "sets of quadruples", where "each space-time point" is identified with such a quadruple (Quine 1976, p. 501). Quine then reflects on this development from physical objects to quadruples of numbers, that is, to abstract objects:

> It is ironical, then, that we at length find ourselves constrained to this antiphysical sort of reduction from the side of physics itself. It is this I have wanted to bring out. Bodies were best, but they needed to be generalized to physical objects for reasons that rested on physical concerns: we wanted

[6] Intensional entities are entities which violate the principle of extensionality: for instance, two concepts can be coextensional without being identical. For details on the concepts of intensionality, intentionality, teleology, and final cause, see below, Sect. 2.1.2.

> to provide designata for mass terms, and we wanted to accommodate physical processes or events. Physical objects, next, evaporated into space-time regions; but this was the outcome of physics itself. Finally the regions went over into pure sets; still, set theory itself was there for no other reason than the need for mathematics as an adjunct to physical theory. The bias is physical first and last, despite the airiness of the ontology. (Quine 1976, pp. 502–503)

This passage evinces the pains that Quine takes to show that his ontological theorizing is driven by the progress of physics. With this ontology of spaciotemporal quadruples, Quine tries to both exclusively quantify over identifiable objects and to provide current developments in microphysics with a suitable ontology. This move, however, comes at a high price. It commits him to a "hyper-Pythagoreanism" (Quine 1976, p. 503) of a sort that only few philosophers dared to defend in modern philosophy—and that must have run counter to Quine's materialist-nominalist predilections. The only things that do exist, according to Quine's hyper-Pythagoreistic proposal, are numbers and quadruples of numbers.

Quine soon abandoned this hyper-Pythagoreistic solution (as can be seen in the passage from *From Stimulus to Science* cited at the beginning of the present section). Instead, he simply accepted the fact that electrons are not identifiable over time: Quine (1974, p. 100) explicitly recommends using the values of quantified variables as objects of reference because this makes it possible to include unspecifiable objects such as electrons. Hence, while Quine clearly had a distaste for non-identifiable objects, this non-identifiability alone does not suffice to exclude a class of things from his ontology, as the case of electrons shows. The fundamental problem with mentalistic meanings is rather that this posit is proposed by a discipline that has no right to posit entities that go beyond the current catalogue of physical entities—only physics can do that.

Compare how Quine (1981 [1964], p. 93) puts this ontological authority of physics in *Smart's Philosophy and Scientific Realism*: "Physics investigates the essential nature of the world, and biology describes a local bump. Psychology, human psychology, describes a bump on the bump." Obviously, a discipline that describes only a bump on a bump is not allowed to ontologically transcend the discipline that investigates "the essential nature of the world" (Quine may have gotten carried away a bit here in his use of essentialist vocabulary).

Quine's heroic struggle with the ontological status of electrons also underscores that Quine is indeed prepared to follow physical theory wherever it goes (even though sometimes not without putting up a fight). Quine refused to clearly specify the parts of his physicalistic ontology. There is no concise list of physical ontological posits to which Quine explicitly committed himself. Rather than to a definitive list of physical entities, Quine commits himself to physics as a discipline.[7]

Physics Is Explanatorily Basic. In addition to establishing that physics determines what is and what is not admissible in ontology, Quine's physicalism also decrees that the explanations of physics are the most fundamental explanations to be had[8]:

> Causal explanations of psychology are to be sought in physiology, of physiology in biology, of biology in chemistry, and of chemistry in physics – in the elementary physical states. (Quine 1979, p. 169)

According to this passage, physics is fundamental in the sense that it provides the most basic level of causal explanation. It makes sense to ask (in a way similar to the habit of children at a certain stage) "Why is that?" when presented with a certain explanation in psychology, for example with regard to the psychologist's claim that her patient is horrified by dogs because he has been bitten by one when he was a small child. This "Why is that?" question could presumably be answered by pointing to some physiological (perhaps neurobiological?) shaping of some structures in his brain. However, it makes no sense to ask "Why is that?" when presented with a certain microphysical explanation for a microphysical

---

[7] Stroud (1990, p. 327) worries that this refusal leaves Quine with an unacceptably vague physicalism, one that cannot even exclude mental entities in principle. Stroud underestimates the power of Quine's physicalism: even if Quine leaves open the exact ontology of a future perfect physics, he is in his rights to say that there is no room for intentional states such as goals in the current ontology of physics. Furthermore, there are not even the slightest signs that would suggest that this could change. However, Stroud is right insofar as this exclusion of intentional states is fallible, because it could in principle be revised by (albeit unlikely) future developments in physics.

[8] I am quoting from Hylton (2007, p. 314).

phenomenon. Explanations of physics are fundamental; there is no way to criticize them from, or reduce them to, an even more fundamental level of consideration.

In addition to being fundamental, physical explanations are also all-encompassing, as the following passage evinces.

> [N]othing happens in the world, not the flicker of a thought, without some redistribution of microphysical states. [...] If the physicist suspected there was any event that did not consist in a redistribution of the elementary states allowed for by his physical theory, he would seek a way of supplementing his theory. Full coverage in this sense is the very business of physics, and only of physics. (Quine 1981 [1978], p. 98)

According to this passage, physics by its very nature aims at full coverage in the sense of being able to account for any event whatsoever. Furthermore, it is the physicist who decides whether current physics needs to be expanded so as to be able to cover a novel kind of event. Furthermore, it would be the physicist who would develop such an expansion of physical theory. Quine (1998, pp. 430–431) is explicit about this final authority of the physicist in the unlikely scenario in which the reality of telepathic events would be established beyond doubt: according to Quine (ibid.) it would be in the authority of physics, and not of psychology, to expand the ontology of physical states such as to accommodate these telepathic events.

This explanatory and ontological hierarchy with physics at the most fundamental level cashes out an idea that was very dear to Quine: the idea of the unity of science. He characteristically used 'science' in the singular.[9] Quine needs a scientific discipline that has the last word on these matters because he rejects the idea of a first philosophy, a non-empirical mode of theorizing prior to science, which could guarantee an explanatory and ontological unity of natural science. Thus, without a discipline that has the last word in ontological and explanatory questions, Quine is endangered to end up with a bewildering number of mutually incompatible ontologies and explanatory approaches.

---

[9] A characteristic example of this Quinean use of 'science' (or, in this case, "the business of scientists") is on Quine (1960, p. 22).

This strong position that Quine accords to physics raises the question whether he was a reductionist of some sort. It is clear that he was no reductionist in the strict semantic or epistemic sense: as a consequence of his verification holism (see below, Sect. 6.1.2), he deemed it impossible to deduce the meaning of or justification for theoretical terms from observational terms. This, however, leaves open the question regarding two kinds of intratheoretical reduction: whether he thought that scientific theories could be translated into the vocabulary of physics, and whether he maintained that the only things that should be assumed to "really" exist are physical entities.

Quine's position on this issue is subtle. On the one hand, as I have shown, Quine accepts that physics has the final authority in ontological and explanatory matters. On the other hand, he held that the semantic reduction of non-physical theories to physical ones is not possible in every instance; thus, Quine (1981 [1978], p. 98) writes that not even all good science could be translated into physics.

Quine's position regarding reduction is best called *eschatological*, since it does not require the reduction to physics in either the ontological or the explanatory dimension right now, but rather is content to prepare everything for the future paradise of an all-encompassing, purely physical theory. This reading is supported by Quine's conception of the status of dispositions, one of the central concepts in his account of first-language acquisition. I will discuss this Quinean concept in more detail below (see Sect. 2.1.3). What matters at present is the ontological and explanatory status of dispositions.

Quine (1974, p. 11) explains the use of dispositions as follows: they are external to scientific theories, playing a regulative rather than a constitutive role, in Kant's understanding of the terms (compare Kant 1998 [1781/1787], B 536ff.). Expressions of dispositions, like solubility in water, can be used to group what looks similar to us and hence may be caused by the same mechanism or physical state. This means that dispositions are only provisional. As it were, they provide scaffolds for the construction of future scientific theories. In this vein, Quine (1974, p. 11) declares that the dispositional idiom has no place in "an ideal language for a finished theory of reality".

This eschatological reductionism allows Quine not only to ban any entity and mode of explanation that has no chance of ever being replaced by proper mechanistic explanations, but also to keep things in place that resist any straightforward reduction to contemporary physics, such as the dispositional idiom.

What reasons did motivate Quine to adopt such a physicalism? Let us start with *On What There Is* (the essay is originally from 1948). At the end of this article, we find Quine cross-pressured between two conceptual schemes (for present purposes, a conceptual scheme can be conceived as an overall theory of the world which includes an ontology). The first conceptual scheme that Quine considers is what he calls the phenomenalistic scheme. It has as values of its bound variables "individual subjective events of sensation or reflection" (Quine 1980 [1948], p. 17). These are what he will later disapprovingly call "sense data". The second conceptual scheme, in contrast, is physicalistic, and it quantifies over "external objects" (ibid.). After urging that simplicity is unable to determine which of these two schemes is preferable, he essentially proposes to give both options a try:

> Here we have two competing conceptual schemes, a phenomenalistic one and a physicalistic one. Which should prevail? Each has its advantages; each has its special simplicity in its own way. Each, I suggest, deserves to be developed. Each may be said, indeed, to be the more fundamental, though in different senses: the one is epistemologically, the other physically fundamental. (Quine 1980 [1948], p. 17)

In this passage, Quine calls the phenomenalistic conceptual scheme epistemologically fundamental. Indeed, on the following page, he declares that, from the phenomenalistic point of view, the ontology of physical objects is a convenient myth: that "myriad scattered sense events come to be associated with single so-called objects" helps us to bring order to the flux of sense data with which we are confronted, but physical objects cannot be deduced from sense data; the sense data alone do not *determine* the precise features of the respective physical objects. Quine does not claim, conversely, that from the point of view of the physicalistic scheme, the ontology of phenomenalistic sense data is a myth. This is an asymmetry that cashes out Quine's claim that the phenomenalistic scheme is epistemologically basic.

There is an obvious difference between this passage and the EN passage (compare above, p. 24). In the passage from *On What There Is*, Quine adopts a pragmatic attitude regarding the choice of conceptual schemes, and he suggests that the phenomenalistic one may be basic for epistemological purposes; in the EN passage, in contrast, he makes it very clear that he intends to use the physicalistic scheme for naturalized epistemology, not the phenomenalistic one.

What is Quine's reason for making this choice? In *Two Dogmas of Empiricism*, he gives the following rationale: "The myth of physical objects is epistemologically superior to most in that it has proved more efficacious than other myths as a device for working a manageable structure into the flux of experience" (Quine 1980 [1951], p. 44). Here, Quine gives a criterion, according to which the physical posits, and with them the physical scheme, excel over its competitors: efficiency in structuring "the flux of experience". In other words, physics deserves this authority because its theories enjoy an unparalleled degree of justification, in Quine's specific sense of empirical-scientific justification, which boils down to prediction of empirical data.

Perhaps there is another reason for Quine's choice. The reason why Quine became fond of the physicalistic scheme is that he saw a chance to have it both ways: by the time of *On What There Is*, being the empiricist that he always was, he felt the need to leave the question open whether the physicalistic scheme (that satisfied his physicalist-ontological desiderata) or the phenomenalistic sense-data scheme (that satisfied his empiricist-epistemological desiderata) should be adopted. However, by the time of *Word and Object*, he thought that by adopting a physicalistic conception of sense data, namely patterns of sensory stimulation of nerve endings, he could pursue the essence of the empiricist project (namely, as the EN passage puts it, to see how evidence relates to theory) within the physicalistic scheme.

To sum up, Quine's physicalism decrees that physics has the final authority in ontological as well as explanatory matters. With regard to ontology, this means that other disciplines are not allowed to posit classes of entities that are clearly beyond the ontology postulated by state of the art physics.

Similarly, explanations from disciplines other than physics should show an openness for future reduction to mechanistic explanations, as Quine's conception of dispositions exemplifies. According to Quine, it is reasonable to give such authority to physics because its theories are extremely well-confirmed, in Quine's sense of empirical-scientific justification. In what follows, I often refer to Quine's entire doctrine of physicalism, and not only to his specific variety of reductionism, as 'eschatological physicalism'.

2.1.2 A Physicalistic Ban on Final Causes

In this section, I detail Quine's rejection of intentionality, intensionality as well as of teleology and final causation. 'Intentionality' is here understood as a technical term, not to be confused with the more commonly used term 'intentional', which means, according to Merriam-Webster, "done by intention or design".[10] 'Intentionality' in its technical use designates the directedness of, among others, mental states and linguistic expressions such as assertions.[11] 'Intensionality' refers to phenomena that violate the principle of extensionality. For instance, two concepts can be co-extensional while still being different concepts.

Final causes are contrasted to efficient causes. An efficient cause is an event or entity that is sufficient for another event or entity to occur and that precedes this event. For instance, the efficient cause for billiard ball number 9's landing in a pocket is that it has been hit by the cue ball (for an overview on the multitude of approaches to cash out this very rough sketch, see Beebee et al. 2009). A final cause, in contrast, typically lies in the future, and it has to do with the aim, goal, or purpose of an action or event. For instance, the final cause for billiard ball number 9's landing in a pocket may be that I have wanted it to land there.

Teleology, finally, can be conceived as the study of the realm of goals, or aims, or purposes. Many questions surround the legitimacy of this study and these concepts for scientific research. As we shall see, Quine shares the popular position that such a study has no place in austere scientific research. Others hold that talk of goals and design might be indispensable

[10]https://www.merriam-webster.com/dictionary/intentional, consulted on May 4, 2017.

[11]Compare Blackburn (2005, pp. 188/127).

to science, especially to biology, but they argue that it should not be taken literally; that is, it should not be assumed that such design and goal really is to be found in nature. According to them, it is just a useful, maybe indispensable jargon (compare Dennett 1987 for a canonical exposition of this view). A third position accepts that specific teleological concepts are indispensable to science and, furthermore, that this reveals something about the nature of reality. Arguably the most prominent contemporary proponent of this view is Tyler Burge (2010).

As we shall see, Quine closely connects intentional phenomena with intensional contexts, final causation and teleology in general—and he rejects the entire group of concepts. As already mentioned, however, it is possible, for instance, to accept teleological language as indispensable and maybe even conducive to scientific progress without drawing any metaphysical conclusions regarding final causes from it.

In *Word and Object*, Quine (1960, pp. 191–232) dedicates an entire chapter to a *Flight from Intension.* Quine (1992, p. 76) understands intensionality as resisting "substitutivity of identity": in intensional contexts, substitution of coreferential terms can change the truth value of the respective statement. This dovetails nicely with the way Quine (1960, pp. 145ff.), more than three decades earlier, introduces expressions of propositional attitudes as causing failures of extensionality by establishing so-called intensional contexts. Consider the construction (1).[12]

1. *Tom believes that Cicero denounced Catiline* (see Quine 1960, p. 145).

Now, 'Tully' is just another name for the person who denounced Catiline, i.e. Cicero. Hence, 'Tully' and 'Cicero' refer to the same being. As a consequence, (2) should follow deductively from (1):

2. *Tom believes that Tully denounced Catiline.*

However, if Tom does not know that Tully is just another name for Cicero, (2) is probably wrong, while (1) is true. Quine (1960, p. 142) calls such

[12]Quine does not use the term 'expression', since it implicates that linguistic structures express something, abstract entities, for instance. He therefore prefers the term 'idiom'. Due to its being an established expression, I continue to use 'expression'.

positions, where substitution by co-referential expressions does not preserve truth value, *non-referential*. Furthermore, Quine (1969a, p. 145) calls the linguistic constructions that are responsible for this non-referentiality, such as expressions of propositional attitudes like 'to believe', *opaque*.[13]

Quine (1960, p. 219) relates his use of 'intentional' to Brentano (and ultimately to scholasticism). Then, he maintains that there is an important division between such intentional idioms and the non-intentional, manageable ones. First, the distinction between non-intentional and intentional expressions marks a division between referential and nonreferential occurrences of terms. Second, it also neatly separates entire ways of conceiving and investigating mental phenomena: non-intentional terms belong to a perspective characterized by behaviorism, efficient causation, and austere theory, while intentional vocabulary belongs to a perspective that makes use of mental concepts, final causation, and non-literal accounts in general.

To see how a verb that does not express a propositional attitude can also be opaque, consider the following example:

3. *Ernest is hunting lions.*

In this example, 'lions', according to Quine (1960, pp. 154f.) can either be fully referential or non-referential. It is referential, for instance, if the sentence is uttered by the director of a hunting reservation that breeds lions for the purpose of being shot by tourists. Then, 'lions' in (3), uttered by the director, is referential. Earnest is then trying to shoot these specific lions that the director knows are out there. However, 'lions' is not used referentially if Earnest is on the hunt in some entirely uncharted territory where, as a matter of fact, there are no lions at all. Then, (3) could still be true, Earnest *could* be hunting lions, although 'lions' in this case does not refer to any actually existing lions. Indeed, even if lions had gone extinct, Earnest could be hunting lions (if he does not know that they are extinct).

Hence, Quine is confronted with the distinction between intentional and nonintentional expressions, which is closely associated with a

[13]Quine (1960, pp. 166–169) proposes an operation of intensional abstraction such that in the formalized version of example (1), the intensional context becomes a reference to an intensional object. However, Quine's ultimate goal is to completely avoid intension.

distinction between behaviorism and mentalism, between efficient and final causes, and between real science and drama (my analysis of pleasure through approval will lend support to this claim). Quine of course opts for behaviorism, efficient causation, and real science, which is why he has to somehow get rid of intentional idioms. Now, what complicates things is that Quine (1960, p. 220) also accepts that "there is no breaking out of the intentional vocabulary by explaining its members in other terms." In other words, he accepts that statements involving intentional verbs, such as 'to believe', cannot be paraphrased without employing other intentional terms. Thus, Quine feels compelled to make a clean sweep and eliminate all intentional talk, at least for serious scientific business:

> One may accept the Brentano thesis either as showing the indispensability of intentional idioms and the importance of an autonomous science of intention, or as showing the baselessness of intentional idioms and the emptiness of a science of intention. My attitude, unlike Brentano's, is the second. (Quine 1960, p. 221)

In this passage, Quine bites the bullet: he grants that intentional explanations are not reducible to non-intentional ones, but then decides to solve the problem by rejecting the entire system of intentional explanations.

Quine's rejection of intentional idioms in this passage implies his rejection of the use of final cause, or purpose, for his explanation of the genesis of successful science (compare again the passage quoted above, p. 24). The same rejection is also implicit in Quine's consistent use of 'mechanism' when he specifies what a *bona fide* explanation should look like—as opposed to the provisional dispositional explanations. Furthermore, Quine rightly fears that allowing final causes for explanatory purposes threatens to bring mentalistic vocabulary with it, since, to act for a purpose, to pursue a certain goal, seems to be inevitably tied to there being a will, as opposed to a mere causal mechanism.

More than three decades later, in *Pursuit of Truth*, Quine (1992, p. 75) conceives the development from final to efficient cause, and the parallel victory of matter over mind as ontological categories of science, as a mark of scientific progress, "vindicated in waves of successful prediction" (of empirical data).

In the next section, I show how Quine tried to avoid any essential use of final causes in his account of the genesis of successful science by employing a version of behaviorism.

2.1.3 Behaviorism: The Physicalist's Psychology

Quine tried to satisfy the constraints imposed by his physicalism by using a specific kind of behavioristic psychology. *The Roots of Reference* (Quine 1974) constitutes his most thorough attempt in this direction. In this section, I delineate Quine's behaviorism mainly based on this work. In an important passage of the book, Quine (1974, p. 4), highlights the potential that he still saw in behaviorist psychology (at a time where mainstream psychology had already abandoned this approach). We find ourselves in a laboratory where an animal is subjected to a specific kind of conditioning: when a screen shows a circle and an animal presses a certain lever, it is rewarded with a pellet of food. When the screen shows four dots arranged in a semicircle and the animal presses the same lever, it is punished for this move by a shock. To test whether it perceives seven spots arranged in a circle as more similar to the circular stripe or to the four dots, we can present the animal with this display and check whether it does press the lever or not. If it presses the lever, it perceives the display as more similar to the circle drawn by a line, if it does not, it perceives the display as more similar to the four dots.

Quine then reflects on this experiment in behavioristic terms. First, he identifies the psychology of learning with the theory of conditioning or of habit formation, of which the experiment just sketched provides an example. Then, he calls the habits that are the consequence of conditioning 'dispositions'. Thus, in the process sketched in the passage, the animal has acquired two dispositions, namely the disposition of pressing the lever when seeing a circle and the disposition of not pressing the lever when seeing the dots.

The way these habits have been acquired is through conditioning: by pleasing the animal after pressing the lever in the first case and by making it feel pain when pressing the lever in the second case.

The controversial claim that underlies Quine's account of first-language acquisition is that this very same process—the conditioning of dispositions, that is, of certain stable, law-like patterns of behavior via pleasure and pain—is an essential part of the explanation of the process of first-language acquisition at least in its early stages.[14] In the following, I detail Quine's concepts of disposition and pleasure.

Disposition. According to Quine, a disposition is used to "refer to a hypothetical state or mechanism that we do not yet understand, or to any of various such states or mechanisms, while merely specifying one of its characteristic effects" (Quine 1974, p. 10). So, a disposition is a way to "earmark" (Quine 1974, p. 11) groups of events that have a characteristic effect and can hence be supposed to share an underlying mechanism. Quine's favorite example of a disposition is solubility in water: for a long time, humans failed to grasp the details of the mechanism that is responsible for a substance's solubility in water. Still, it was not objectionable to use the dispositional idiom 'soluble in water' even in the absence of knowing the mechanism if it was kept in mind that one had rather earmarked a group of superficially similar phenomena than provided a genuine explanation.

Furthermore, Quine wants to conceive of a fully competent speaker's verbal behavior as a bundle of dispositions as well (Quine 1974, p. 15). Thus, Quine claims that there is a specific physical mechanism underlying each specific sentence that a subject is able to produce in correct circumstances. Quine (ibid.) argues that this is plausible since the dispositions can share parts. Note that this is an almost completely deterministic view of language ('almost' because our need to use the dispositional idiom introduces a *ceteris paribus*): given a certain trigger, the subject will almost inevitably utter a given sentence, just like salt will almost inevitably dissolve in water. A pertinent question with regard to dispositions to verbal behavior is: what are the specific triggers for the utterance of a certain sentence in a certain situation? Quine attributes this question to Chomsky

[14]Compare Quine (1960, pp. 33–34) and Quine (1995, p. 21).

and asserts that it can be answered with recourse to "query and assent" (Quine 1974, p. 15).[15]

Pleasure. Quine gives the following behavioristic explanation of his concept of pleasure:

> To learn is to learn to have fun. Behaviorally, the shoe is on the other foot: an episode counts as pleasant if, through whatever unidentified mechanism of nerves and hormones, it implants a drive to reproduce it. The pleasure is measured by the strength of this drive. (Quine 1974, pp. 28–29)

According to this passage, whether or not a subject experiences an episode as pleasant is determined by observing its drive to reproduce the episode. Again, Quine assumes that there is an underlying mechanism that determines this behavior, but this time, it is cast in physiological, not in physicalistic vocabulary: he speculates that it may involve nerves and hormones.

It will become clear later that this basic conceptual inventory avoids any essential use of final causes: the animal is never credited with purposive action, say, to get a pellet of food or to avoid a shock. Rather, it is caused to act in these ways by the conditioning to which it has been subjected: if a certain type of response to a certain type of stimulus results in a pleasant experience for the animal, it acquires the disposition to act again in the same way when stimulated similarly; the disposition *causes* the animal to act in one way or another. The same holds for Quine's conception of language itself: it is free of any communicative purposes, just a complicated bundle of dispositions to behavior.[16]

This austerely behavioristic approach to language learning dovetails nicely with Quine's eschatological physicalism: Quine's central concept of disposition prepares the ground for a fully-fledged mechanistic explanation. Furthermore, it ties well with Quine's empiricist distaste for innate

---

[15]For a thorough comparison of Chomsky's and Quine's views on language, compare Smith (2014).

[16]Hylton (2007, p. 106) acknowledges this: "Quine sees language as embodied in dispositions to behavior. Actual behavior is the *evidence* for someone's possession of the dispositions which constitute a mastery of a given language. But the underlying reality for which it is evidence consists of *dispositions* to behavior, and these are enduring (though changeable) states of the organism." Compare also Hylton (2016, p. 114).

cognitive structure, which will be the topic of the following section. Note that I will, like Quine,[17] sometimes use intentional vocabulary to describe the infant's acquisition of her first observation sentences; this is done for the sake of brevity and clarity. The real explanation, however, is always in terms of dispositions, stimuli, and conditioned responses.

2.1.4 Source- and Checkpoint-Empiricism

Quine's empiricism contains two different aspects, which can be called *source-empiricism* and *checkpoint-empiricism* (below, Sect. 6.2.2, I discuss a passage where Quine actually draws the distinction, see also Quine 1992, p. 21). Quine explains source-empiricism in the following passage:

> The most notable norm of naturalized epistemology actually coincides with that of traditional epistemology. It is simply the watchword of empiricism: *nihil in mente quod non prius in* sensu. This is a prime specimen of naturalized epistemology, for it is a finding of natural science itself, however fallible, that our information about the world comes only through impacts on our sensory receptors. (Quine 1992, p. 19)

In this passage, Quine declares that the most important norm of naturalized epistemology coincides with the one of traditional epistemology, namely the principle that nothing is in the mind, no piece of knowledge, no sentence, that has not been in the senses before. In accordance with his physicalism, Quine declares that the relevant information must come through "impacts on our sensory receptors", that is, stimulations of our nerve endings. Quine claims that empiricism, thus conceived, is a scientific insight, and hence a prime specimen of naturalized epistemology. I call this doctrine, according to which our only source of information about the world is given by patterns of sensory stimulation, *source-empiricism.*

For Quine's account of first-language acquisition, this means that, with the notable exception of innate quality spaces, or senses of similarity, or similarity standards (Quine uses all three terms for essentially the same innate structure, see below, Sects. 2.2.1 and 3.1.2), the mind of every

---

[17] Compare Quine (1974, p. 33).

infant is a *tabula rasa*, a blank slate, when it first sees the light of the world; there are no innate concepts or almost conceptual structures, since they could not come to the infant's mind through the senses.

Furthermore, these raw, unprocessed patterns of sensory stimulations are ontogenetically and epistemically basic: these patterns *are* all the information that the infant has available to learn her first language, and they *are* all the evidence the scientist has to test her theories. In contrast, Johnsen (2014, pp. 983f.) holds that these patterns of sensory stimulations do not play such a central role: According to Johnson (ibid.), Quine only maintains that an individual's sensory organs have no other "inputs" than patterns of sensory stimulations (Johnsen 2014, p. 984). However, Johnsen argues, this does not rule out that the infant itself has much more elaborate information and evidence available, from, say perception. For instance, one could assume that the human visual system, as it were, automatically parses the proximal stimulus into representations as of discrete, bounded objects (a view along these lines is Burge's, which will be discussed below, Sect. 4.2).

My strict reading is supported by my discussion of Quine's conception of stimulus meaning (see below, Sect. 2.2).[18] Further support for my reading comes from the fact that Quine was deeply troubled to find that nerve endings are non-homologous[19] (compare below, Sect. 2.2.2). If raw, unprocessed patterns of sensory stimulations would not be ontogenetically and epistemically basic, this finding should not have troubled Quine at all. Finally, as I will argue below (Sect. 6.1), Quine needs a conception of evidence that is unstructured enough to underpin his verification holism, which in turn justifies his empiricist justificatory monism. The kind of evidence that Johnsen seems to have in mind therefore threatens to undermine Quine's justificatory holism and ultimately his naturalism.

Source-empiricism, the claim that our only source of information about the world is given by patterns of sensory stimulation, is different from the

[18]Compare also Glock (2013, p. 581), who shares my view, as he maintains that "Quine grants that observation sentences are, by-and-large, *about* macroscopic objects [...]. At the same time, he insists that the evidence on which speakers assent to observation sentences consists of neural triggerings."

[19]In this context, 'homology' means having the same, or at least a very similar structure. Hence, the nerve endings of two humans are homologous if they have the same structure. This requires that their sensory surfaces are isomorphic.

claim that predictions of such patterns of sensory stimulations also constitute the ultimate checkpoints for science. This is *checkpoint-empiricism*, which will be addressed below (compare Sects. 6.1.3 and 6.2.2 and Quine 1992, p. 20). Quine (1992, p. 21) rightly argues that one can adopt checkpoint-empiricism and reject source-empiricism. Quine's example is telepathy: it is conceivable that somebody admits telepathic information for the construction of scientific theories, but rejects them as checkpoints for these theories.

It will become clear in the further course of this chapter that checkpoint-empiricism is immaterial to Quine's account of first-language acquisition. This is because the infant is not actively testing any hypotheses; rather she is caused to act in certain ways in response to certain stimuli. There is no room for checking hypotheses in Quine's thoroughly behavioristic approach to first-language acquisition.[20]

To conclude this section, note that Quine's physicalism, his rejection of intentionality and of final causes, his behaviorism, and his source-empiricism nicely support each other—while none of them directly implies any of the other. Eschatological physicalism prohibits attributing to the infant any mental, intentional states to explain first-language acquisition. Hence, the rejection of final causes, behaviorism as an overall methodology and Quinean source-empiricism are congenial to the eschatological physicalist. If you are a behaviorist, in turn, one of the main motivations for adopting this psychological methodology is some version of physicalism, or at least a strong aversion against intentional entities; again, you will find Quine's empiricism very appealing. Finally, if you accept Quinean source-empiricism, you probably are already an eschatological physicalist.

---

[20]There seems to be a problem for Quine here (compare below, Sect. 3.3.2): if mastering a language is having a primitive scientific theory, and if mastering a language has nothing to do with deliberately checking hypotheses, how then can scientific theorizing be based on checking hypotheses? As Marco Toscano has pointed out to me, this might be the problem that John McDowell raises for Quine in the Appendix to *Mind and World* (McDowell 1994).

2.2 The Elegantly Physicalistic, But Flawed Account of *Word and Object*

In this section, I first delineate Quine's account of how observation sentences relate people both to each other and to the world, as he gives it in *Word and Object* (Sect. 2.2.1; at one point, I am also referring to *Pursuit of Truth*). This account makes extensive use of the notion of intersubjectively invariant stimulus meaning. Then, I show how Quine himself came to accept that it is factually wrong to assume such intersubjectively invariant stimulus meaning (Sect. 2.2.2).

2.2.1 The Account of *Word and Object*

So, how does an infant acquire her first observation sentences? Quine gives an overview on the process in the following passage:

> An oddity of our garrulous species is the babbling period of late infancy. This random vocal behavior affords parents continual opportunities for reinforcing such chance utterances as they see fit; and so the rudiments of speech are handed down. The babbling is a case of what Skinner calls *operant behavior*, "emitted rather than elicited." Operant behavior can be selectively reinforced, in people and other animals, by quick reward. The creature tends to repeat the rewarded act when stimuli recur that chanced to be present at the original performance. What had been stimulation accidentally attending the act becomes transmuted, by the reward, into a stimulus *for the act.*
>
> The operant act may be the random babbling of something like 'Mama' at some moment when, by coincidence, the mother's face is looming. The mother, pleased at being named, rewards this random act, and so in the future the approach of the mother's face succeeds as a stimulus for further utterances of 'Mama'. The child has learned an occasion sentence. (Quine 1960, pp. 80–81)

According to this passage, it all begins with "the babbling period of late infancy". This is a period in which the infant randomly utters word-like noises. By mere chance, Quine holds, the child will utter something reminiscent of an appropriate observation sentence during this babbling period. For example, it could utter something like "Mama!" when the parent's face is in sight. If the parent hears this utterance, she will certainly reward this random utterance. Even if it is not explicit in this passage (but see the detailed discussion of the process as described in *The Roots of Reference* below, Sect. 3.3.2), this reward, or several of them, will cause the infant to acquire a disposition to utter the same sound whenever being stimulated in the same way. The acquisition of this disposition means, on Quine's conception, that the infant has learned her first sentence: it utters the sentence in the appropriate situations. Note that, according to this conception, the infant who has learned her first observation sentence has not understood anything regarding the meaning, traditionally understood, of 'Mama'.

In the remainder of this section, I detail this process by discussing Quine's concepts of observation sentence, stimulus meaning, and quality space.

According to Quine, the infant's first steps into language are not the acquisition of words, but rather of observation sentences. In *Word and Object*, Quine (1960, pp. 35–46) introduces observation sentences as a variant of occasion sentences. The latter "command assent or dissent only if queried after an appropriate prompting stimulation" (Quine 1960, pp. 35–36). In this, they differ from standing sentences such as "Physical bodies attract each other". Occasion sentences are easier to learn, as a competent speaker's assent to them depends—to a considerable degree—on the respective situation.

Furthermore in most cases, an observation sentence has two substantially different modes of existence, or more precisely, of use. First, it can be used as a *bona fide* structured sentence. In this mode, observation sentences have a grammatical and logical structure and they carry ontological commitments—they are part of a theory of the world. Second, in addition to this conceptually structured and thus theory-laden use, observation sentences have a holophrastic, non-conceptual use (Quine 1992, p. 8). In this mode, which is the epistemically significant one, observation sentences are meaningful only as a whole, as if they were composed of a single word.[21]

[21]Thus Quine's use of the linguistic term holophrastic, which normally refers to the one-word sentences a child produces at an early stage of language learning.

Hence, they do not have any (proto-)logical structure, and they are not committed to any ontology. In the following, if not specified otherwise, with 'observation sentence', I always mean their holophrastic use.[22]

That observation sentences can be used holophrastically, and hence theory-freely, constitutes the defensive part of Quine's explanation of how observation sentences can connect us to the world. Holophrastic observation sentences allow for a view on the world which is not blurred by one's own language, or science. This qualifies holophrastic observation sentences as theory-neutral checkpoints for research. When it comes to holophrastic observation sentences, it is nature, not man, which gives the verdict on an observation sentence in a given situation. Furthermore, this same independence from language and theory qualifies observation sentences as the port of entry into language for an infant with no prior knowledge of language. Observation sentences are precisely those sentences which can be learned without any prior knowledge of the language in question.

The positive part of Quine's theory of observation sentences—the one that explains the possibility of making contact with the world—is based on his notion of *stimulus meaning*. Often, the meaning of a word is conceived as something that is not located in spacetime. For example, the meaning of the word 'horse' could be said to denote an abstract property which is common to all horses. Quine never accepted such a non-physicalistic conception of meaning. He therefore tried to reconceive the notion of meaning in a way compatible with his overall philosophical outlook. Stimulus meaning is what Quine proposes as a physicalistic reconception of the traditional notion of meaning.

Quine (1960, p. 32) defines the *stimulus meaning* of a sentence P relative to the subject S in the following way. The *affirmative* stimulus meaning of P for S is the class of all the sensory stimulations—across all the subject's senses—which prompt S to assent to P. *Mutatis mutandis*, Quine defines the *negative* stimulus meaning of P for S: it is given by the class of stimulations that cause S to reject P, and the stimulus meaning of P

[22]The examples that I will use frequently, like "Red!", do rather suggest a holophrastic use. This, however, is the exception rather than the rule. I use it nonetheless in the hope that emphasizing the holophrastic aspect of observation sentences by the use of such examples proves helpful in presenting Quine's theory.

emerges as the ordered pair of the two classes.[23] To illustrate this notion of stimulus meaning, consider the sentence "Rabbit!". What is the stimulus meaning of this sentence for our speaker Rudolph? According to the definition above, it consists of the class of the stimulations which would cause Rudolph to affirm the sentence together with the class of stimulations which would cause him to deny it.[24]

Quine (1960, pp. 44–45) introduces the notion of *observationality* to set the stimulus meaning of observation sentences apart from the stimulus meaning of other sentences. A sentence is low on observationality if the intersubjective variability of its stimulus meaning is high. Consider the following two examples. On the one hand, the observationality of "A bachelor!" is supposedly very low, as your inclination to assent to it simply depends on whether you know the person in question or not—and not on some observable feature of the respective scene (putting aside things like wedding rings, or their absence). Thus, the stimulus meaning of "A bachelor!" differs among people according to their different social relationships. On the other hand, the observationality of "Red!" is very high, since the stimulus meaning of it is quite uniform among competent English speakers: according to *Word and Object*, we are all prepared to agree to this latter sentence if, and only if, our retina is stimulated at a certain wavelength. In short, *the stimulus meaning of a given observation sentence is intersubjectively almost invariant.*

Since observation sentences, taken holophrastically, are directly associated with an intersubjectively invariant stimulus meaning, to judge whether a given observation sentence is true in a given situation does not presuppose any linguistic knowledge. As a consequence, the infant has something that it shares with her parent before she can rely on a

---

[23]Quine gives the final definition of stimulus meaning as follows: "a stimulus meaning is the stimulus meaning *modulo n* seconds of sentence S for speaker *a* at time *t*" (Quine 1960, p. 33). The specification of the modulus and the time are required to avoid some technical problems. For my purposes, the definition without the modulo is sufficient.

[24]Føllesdal (2014, p. 268) worries that taking sensory stimulations as basic semantic units of significance contradicts Quine's behaviorism, since "they are not part of the publicly available evidence that is so crucial to behaviorism". I think this worry can be met (to the extent to which it can be met, see Sect. 2.2.2) in a rather simple way: just take the subject into a laboratory, where behaviorism is naturally at home, and wire her up such that the researchers can monitor all the activities on the subject's nerve endings. This furnishes all the public availability that Quine's behaviorism requires.

shared language, and which can *ipso facto* serve as the gateway to language. Furthermore, thanks to the intersubjective invariability and theory-freedom of the stimulus meanings of observation sentences, even proponents of radically different theories will affirm these observation sentences in the same situations, that is, when stimulated similarly in the relevant respect. Thus, holophrastic observation sentences can provide theory-neutral checkpoints for scientific theorizing.

Given this conception of stimulus meaning, it is evident that Quine here presents an externalist perspective on the meaning of observation sentences and thus on empirical evidence. While introspection will never reveal to a subject the true meaning of her observation sentences, physicists can specify once and for all what "Red!", "Lightning!" etc. mean, according to this Quinean picture.

However, even if we grant that observation sentences, used holophrastically, do not presuppose linguistic knowledge, or a theory of the world, this does not adequately explain how they can function as the gateway to language. Regarding the sheer multiplicity of causal patterns of sensory stimulations which accompany any of the parent's utterings of "Red!"—sounds, smells, visual patterns, etc.—it is highly unlikely that the child is able to identify the important aspect of this torrential input, namely the presence of a red surface. Quine addresses this problem with his conception of innate quality spaces, or innate senses of similarity, or innate similarity standards (Quine uses these three expressions for essentially the same innate structure). To learn her first observation sentences, Quine holds, the child must be inclined to give more weight to the relevant similarities than to the irrelevant ones—where the relevant similarities are those which matter for the meanings of the observation sentences that the infant tries to learn:

> If the child is to be amenable to such training, however, what he must have is a prior tendency to weight qualitative differences unequally. He must, so to speak, sense more resemblances between some stimulations than between others. Otherwise a dozen reinforcements of his response 'Red', on occasions where red things were presented, would no more encourage the

> same response to a thirteenth red thing than to a blue one; [...] In effect therefore we must credit the child with a sort of pre-linguistic quality space. (Quine 1960, p. 83)

According to this passage, the child is biased to generalize along the aspects of the stimuli which matter for her parent, namely the red patch in the visual field, as opposed to, say, a gust of wind. The reason for this bias is that the infant's quality spaces situate global stimuli involving red close together, instead of grouping all those global stimuli where a certain distinctive gust of wind is present. This way, the child is able to inductively generalize the stimulus meaning of her first observation sentences. If we remember that the stimulus meaning of observation sentences is *uniform* among the speaking community, it follows that, by having thus learned these observation sentences, the child has learned the meaning of her first words.[25]

The innate quality space, which Quine attributes to the child to explain how it learns her first observation sentence, is of central importance for Quine's account of first-language acquisition.[26] Furthermore, observation sentences, which are, as it were, prefigured in the infant's quality space, constitute the ultimate checkpoints for science. Therefore, it depends on the properties of this quality space whether natural science is able to, as Quine (1974, p. 19) puts it with deliberate vagueness, "run with the grain of nature". I elaborate on this significance of Quine's conception of quality space below (Sect. 3.1.1).

It is remarkable that Quine writes of the child sensing more resemblance between some stimulations than between others. In the passage quoted above, Quine hedges this way of putting it. However, later on, Quine (1960, p. 84) explicitly maintains that stimulations are the kind of things that are organized in quality spaces.

---

[25]Note again that the intentional vocabulary is here merely to facilitate presentation. The account to which Quine ultimately commits himself is phrased in terms of efficient causes, dispositions, stimuli, and conditioned responses, see Sects. 2.1.3 and 3.3.2.

[26]There is surprisingly little research on Quine's notion of quality space. Exceptions are Broughton (1981), who suspects that Quine's notion is covertly mentalistic, and van Brakel (2000), who compares Quine's notion to similar conceptions in cognitive science.

In summary, this yields a rather neat and physicalistic picture of how the infant makes her first steps into language: having uttered at random something along the lines of "Red!" in the presence of a red surface, the parent rewards the infant. As the infant likes being rewarded, she repeats the same sound on occasions when she is stimulated in a similar way, which means that she has started using the observation sentence "Red!" correctly. Since the possible respects in which another stimulation could be similar to the relevant one are legion, the child needs a quality space that is both biased towards the relevant similarities and similar to the one of her parent. Thanks to this quality space, the infant can associate the correct stimulus meaning with "Red!", a stimulus meaning which is intersubjectively (almost) invariant. Hence, on this Quinean account, for the infant to have learned the observation sentence "Red!" means nothing else than that the infant has been successfully conditioned to respond in a certain way—by uttering "Red!"—to certain patterns of stimulations, namely the stimulus meaning of "Red!".

2.2.2 The Flaw: Stimulus Meanings Are Private

In *Word and Object*, Quine maintains that observation sentences, conceived holophrastically, have a very similar stimulus meaning *across* different speakers. This intersubjective similarity is a central tenet of Quine's account of how the infant learns her first observation sentences. However, in *From Stimulus to Science*, Quine explicitly declares that "global stimuli are private" (Quine 1995, p. 20). What happened?

The first paper in which Quine explicitly acknowledges the fact that global stimuli are private—and recognizes its relevance for his theory of meaning and evidence—is *Propositional Objects*. Quine (1969b, p. viii) writes that he held *Propositional Objects* as a lecture in 1965. Thus, the following remarkable passage probably dates from no more than five years after the publication of *Word and Object*:

> It seems vital that in correlating one subject's verbal behavior with another's, for instance as a basis for translating one language into another, we be able to equate one subject's stimulation to another's. Yet how are we to do so? If we construe stimulation patterns my way, we cannot equate them without

supposing homology of receptors; and this is absurd, not only because full homology is implausible, but because it surely ought not to matter. (Quine 1969b, p. 157)

Here, Quine explicitly acknowledges that his notion of stimulus meaning and thus of communication in general depends on homology of sensory receptors, which is implausible.[27] Why is this the case?

Homology of sensory receptors, which is a condition for the possibility of intersubjectively shared stimulus meaning, is not required when the focus is restricted to one single subject. When asked for a verdict on an observation sentence in two different situations, a competent speaker behaves similarly when stimulated in a similar way: she assents to it when confronted with a stimulation which falls in the range of the sentence's affirmative stimulus meaning, and she dissents when the stimulation belongs to the sentence's negative stimulus meaning. However, in spite of the claim of *Word and Object*, these stimulus meanings are private. This means that two witnesses of the same scenario who agree on the truth of an observation sentence cannot be said to witness the same situation *sensu stricto*, that is, they cannot be said to be stimulated alike. In *Propositional Objects*, Quine first states that, in the context of first-language acquisition, sensory surfaces are the best "boundaries" to define input, but that "when we come to the seemingly essential business of saying what it is for two people to be stimulated alike, we tangle with the myth of homologous nerve endings" (Quine 1969b, p. 158).

Quine here wrestles with the following issue: if, due to the lack of homology of nerve endings, the parent's global stimulus differs from the child's in any given situation—even if the child could take the very same physical position as her parent—then it is misguided to explain the success of the child's inductions by means of a quality space which situates global stimuli. No matter how biased her quality space, the child can never hope to infer a stimulus meaning which is substantially similar to the parent's because no two speakers associate even approximately similar stimuli with a

[27]This passage resounds in *Pursuit of Truth*, where Quine claims that he felt discomfort over the notion of sameness of stimulus meanings already in 1965, and he expresses his conviction that "anatomical minutiae" like homology of nerve endings should be irrelevant to his project, see Quine (1992, pp. 3–4, 40).

given observation sentence. In other words, even if there were a remarkable similarity between the quality spaces of the parent and the infant, it would still be useless because there are no similar global stimuli to situate in these spaces. Quine puts this dependence of his notion of communication on intersubjectively shared global stimuli as follows:

> It is when we turn to language theory that the problem [of equating one subject's stimulation to another's, RG] becomes acute. Being social, language depends on associating utterances with stimulations that can be publicly identified in their recurrences from occasion to occasion and speaker to speaker. (Quine 1969b, p. 158)

A simple way to bypass this problem is unavailable to Quine. One could simply accept that global proximal stimuli are private, but assume the distal stimuli to be relevant for the purposes of naturalized epistemology, that is, the intersubjectively invariant objects that cause the proximal stimulations. Thus, the child could be said to learn the meaning of "Red!" by perceiving the red thing which the parent perceives too and to which the parent refers by uttering "Red!".

Quine, however, does not think that such a strategy is available to him: "It is the stimulation at the bodily surface that counts, and not just the objective existence of objects of reference off in the distance" (Quine 1969b, p. 158). Why is it the stimulation of sensory surface that counts for Quine's purposes, and not the corresponding distal stimulus? Because Quine (1992, p. 41) is "interested in the flow of evidence from the triggering of the senses to the pronouncements of science". Assuming that infants are able to refer to objects, if only perceptually, before they even learn their first observation sentence would omit the part of naturalized epistemology in which Quine is highly interested: how did humans develop this ability to refer to objects, and how is it justified? It therefore runs counter to the very goal of Quine's project to assume the distal stimulus as basic for first-language acquisition.

Furthermore, presupposing that the infant has this ability to refer to objects from the start stands in tension with Quine's eschatological physicalism and his source-empiricism. His physicalism requires that explanations in psychology be open to future mechanistic reduction; but it would be very difficult to achieve such a reduction if the explanation would simply assume that infants have this intentional ability from the start. Of course, such innate abilities are incompatible with Quine's source-empiricism.

Similarly, his behavioristic psychology forbids him to credit humans with any abilities before they are observable in behavior. Objective reference, Quine holds, is only observable in sophisticated verbal behavior.

Note that the present issue is not ontological, but epistemico-psychological. The question is not whether we should accept an ontology that includes physical objects, but rather whether we should assume that infants are able to perceive physical objects quite some time before having acquired the ability to speak about them to some degree of sophistication.

In effect, at the end of *Propositional Objects*, Quine admits that he has no solution to this problem. Regarding the homology problem, he remarks: "I leave you, therefore, with a problem of theoretical formulation that carries no practical problem with it" (Quine 1969b, pp. 159–160). However, as I hope to show in the following, Quine was far from simply leaving this problem to his readers. On the very contrary: he grapples with it for the rest of his publishing career.

References

Beebee, Helen, Christopher Hitchcock, and Peter Menzies. 2009. Introduction. In *The Oxford Handbook of Causation*, ed. Christopher Hitchcock, Helen Beebee, and Peter Menzies, 1–20. New York: Oxford University Press.

Blackburn, Simon. 2005. *The Oxford Dictionary of Philosophy*, 2nd ed. Oxford: Oxford University Press.

Broughton, Lynne M. 1981. Quine's Quality Space. *Dialectica* 35: 291–302.

Burge, Tyler. 2010. *Origins of Objectivity*. Oxford: Oxford University Press.

Dennett, Daniel C. 1987. *The Intentional Stance*. Cambridge: The MIT Press.

Føllesdal, Dagfinn. 2014. Developments in Quine's Behaviorism. In *A Companion to W.V.O. Quine*, ed. Gilbert Harman and Ernie Lepore, 263–278. Chichester: Wiley Blackwell.

Gibson, Roger F. 1988. *Enlightened Empiricism. An Examination of W. V. Quine's Theory of Knowledge*. Tampa: University of South Florida Press.

Glock, Hans-Johann. 2013. Quine and Davidson. In *A Companion to Davidson*, ed. Ernie Lepore and Kirk Ludwig, 567–588. Chichester: Wiley Blackwell.

Hylton, Peter. 2007. *Quine*. New York: Routledge.

———. 2016. Quine's Philosophies of Language. In *Quine and His Place in History*, ed. Frederique Janssen-Lauret and Gary Kemp, 103–122. London: Palgrave Macmillan.

Johnsen, Bredo C. 2014. Reclaiming Quine's Epistemology. *Synthese* 191: 961–988.

Kant, Immanuel. 1998 [1781/1787]. *Kritik der reinen Vernunft*, ed. Jens Timmermann. Hamburg: Meiner.

McDowell, John. 1994. *Mind and World*. Cambridge: Harvard University Press.

Quine, Willard Van Orman. 1960. *Word and Object*. Cambridge: The MIT Press.

———. 1969 [1958]. Speaking of Objects. In *Ontological Relativity and Other Essays*, 1–25. New York: Columbia University Press.

———. 1969a. Epistemology Naturalized. In *Ontological Relativity and Other Essays*, 69–90. New York: Columbia University Press.

———. 1969b. Propositional Objects. In *Ontological Relativity and Other Essays*, 139–160. New York: Columbia University Press.

———. 1974. *The Roots of Reference*. LaSalle: Open Court Publishing.

———. 1976. Whither Physical Objects? In *Boston Studies in the Philosophy of Science*, vol. XXXIX, 497–504. Dordrecht: Reidel.

———. 1979. Facts of the Matter. In *Essays on the Philosophy of W.V. Quine*, ed. R. W. Shahan and C. Swoyer, 155–169. Hassocks: Harvester Press.

———. 1980 [1948]. On What There Is. In *From a Logical Point of View*, 1–19. Cambridge: Harvard University Press.

———. 1980 [1951]. Two Dogmas of Empiricism. In *From a Logical Point of View*, 20–46. Cambridge: Harvard University Press.

———. 1981. On the Very Idea of a Third Dogma. In *Theories and Things*, 38–42. Cambridge: Harvard University Press.

———. 1981 [1964]. Smart's Philosophy and Scientific Realism. In *Theories and Things*, 92–95. Cambridge: Harvard University Press.

———. 1981 [1978]. Goodman's Ways of Worldmaking. In *Theories and Things*, 96–99. Cambridge: Harvard University Press.

———. 1992. *Pursuit of Truth*, 2nd ed. Cambridge: Harvard University Press.

———. 1995. *From Stimulus to Science*. Cambridge: Harvard University Press.

———. 1998. Reply to Hilary Putnam. In *The Philosophy of W. V. Quine. Expanded Edition*, ed. Lewis Edwin Hahn and Paul Arthur Schilpp, 2nd ed., 427–431. LaSalle: Open Court Publishing.

Smith, Barry C. 2014. Quine and Chomsky on the Ins and Outs of Language. In *A Companion to W.V.O. Quine*, ed. Gilbert Harman and Ernie Lepore, 483–507. Chichester: Wiley Blackwell.

Stroud, Barry. 1990. Quine's Physicalism. In *Perspectives on Quine*, ed. Robert B. Barrett and Roger F. Gibson, 321–333. Cambridge: Blackwell.

van Brakel, Jaap. 2000. Quine and Innate Similarity Spaces. In *Quine: Naturalized Epistemology, Perceptual Knowledge and Ontology*, ed. Lieven Decock and Leon Horsten, 81–99. Poznan Studies in the Philosophy of the Sciences and the Humanities, 70. Amsterdam: Rodopi.

3

Quine II: The Evolutionary-Perceptual Account

In this chapter, I first trace Quine's attempts, stretching over almost four decades, to eliminate the flaw with which the last chapter ended (Sect. 3.1). Briefly, Quine kept the basic layout of the account given in *Word and Object* intact; in particular he insisted that it is the observation sentence that plays the key role in connecting language to the world and humans to each other. What changed, however, was that perceptual psychology as well as evolutionary psychology came to play a much larger role in his thinking, at the expense of his physicalism and empiricism. It will turn out that Quine's attempts focused on finding a suitable replacement for intersubjectively invariant stimulus meanings. Then, after summarizing Quine's journey (Sect. 3.2), I criticize two central aspects of his mature account (Sect. 3.3).

3.1 Quine's Struggle to Explain Intersubjectivity

In this section, I trace Quine's attempts to solve a problem that he recognized around 1965: his conception of intersubjectively (almost) invariant

R. Gubelmann, *A Science-Based Critique of Epistemological Naturalism in Quine's Tradition*, https://doi.org/10.1007/978-3-030-24524-5_3

stimulus meaning is based on a factually wrong presupposition, namely on homology of nerve endings. Since the nerve endings of different humans are non-homologous, the stimulus meanings of any observation sentence will *inevitably* differ from speaker to speaker: differently structured nerve endings cannot be stimulated in the same pattern. Hence, stimulus meanings are *private*, as Quine (1992, p. 44) puts it. This means that speakers of the same language, including scientists, are necessarily talking past each other even in the case of observation sentences, which undermines the possibility of intersubjectivity and hence the very possibility of communication. This in turn makes it impossible for an infant to acquire her first language.[1]

Along with this main topic, I delineate the evolution of Quine's answer to another question: why is our science so predictively successful? Since science is epistemically based on observation sentences, and since these observation sentences are in turn shaped by our innate quality space, an answer to this question must involve attributing some sort of epistemic validity to these quality spaces.

In the following sections, I show how Quine tried to come to terms with these issues in *The Roots of Reference* (Sect. 3.1.2), *Pursuit of Truth* (Sect. 3.1.3) and *From Stimulus to Science* (Sect. 3.1.4). This will show that Quine's final proposal to address the two issues is not recognizably physicalistic anymore, but at best biological, and that it makes ample use of innate cognitive structures, which is hard to reconcile with empiricism. Before doing that, I discuss Quine's rather surprising position in *Natural Kinds* (Sect. 3.1.1).

3.1.1 Natural Kinds: A Transitory Period

Before considering Quine's positive proposals concerning the explanation of intersubjectivity in spite of private stimulus meanings, I show that the problem is right beyond the surface in Quine's article *Natural Kinds* (first

[1] I am here not referring to the indeterminacy of translation. The essential point here is that Quine cannot explain communication at a very basic level, a level that Quine did not want to be touched by any indeterminacy, namely the level of holophrastic observation sentences. Quine intended holophrastic observation sentences to furnish true intersubjectivity, almost entirely unaffected by indeterminacies.

presented 1967, published 1969). The article also serves to make explicit the epistemic dimension of quality spaces.

Here is how Quine conceives the central function of quality spaces, and their inhabitants, in *Natural Kinds*: "The uniformity of people's quality spaces virtually assures that similar *presentations* will elicit similar verdicts [emphasis RG]" (Quine 1969, p. 125). When compared to *Word and Object* this passage conceives of the function of quality spaces in the same way: to guarantee that different humans assent to observation sentences in roughly the same situations. However, in contrast to *Word and Object*, the inhabitants of quality spaces are now vaguely referred to as 'presentations'.

In light of the privacy of stimulus meanings, it is clear why Quine uses 'presentations' instead of 'stimulations': for the reasons presented above (Sect. 2.2.2), a quality space that allocates inevitably different stimulations and hence stimulus meanings cannot ensure intersubjectivity of meaning, no matter how uniform the space itself is. If global stimuli are necessarily different from speaker to speaker, then stimulus meanings, being a part of global stimuli, are necessarily different too, no matter how uniform the respective similarity spaces are. As it were, if you take an apple and an orange and put them into two identical storage systems, the two fruits are still not identical, not shared across the two storage systems. This impasse notwithstanding, there are still passages in *Natural Kinds* where Quine (1969, p. 123) refers to an "innate spacing of stimulations" which is at work in first-language acquisition. I submit that this ambiguity in use is indicative of a transitory phase in Quine's theory.

Apart from this uncertainty about the ontological level on which similarity standards operate, Quine suggests that there is not one but at least two parallel standards of similarity at work in each human. First, there is an innate sense of similarity—or quality space, Quine uses both terms interchangeably in *Natural Kinds*—that is based on colors and that thus proves very useful "at the food-gathering level" (Quine 1969, p. 127). It is also this *innate* sense of similarity which is vital for first-language acquisition. Unfortunately, "color is king in our innate quality space, but undistinguished in the cosmic circles" (Quine 1969, p. 127). In other words, our innate sense of similarity is useful to survive, but not of much use in understanding the laws behind the experiences. In Quine's words: it is a bad guide to the "functionally relevant groupings in nature" (Quine

1992, p. 126). This, however, is not the end of the story, as natural selection has been particularly generous towards human beings:

> Evidently natural selection has dealt with the conflict by endowing man doubly: with both a color slanted quality space and the ingenuity to rise above it. (Quine 1969, p. 128)

Here, Quine proposes the following model to explain the capacities of human beings to both succeed on the "food-gathering level" (Quine 1969, p. 127), that is, in their daily affairs, and in theoretical science. When it comes to the former, we are well-off with our inborn quality space. When it comes to exploring the "workings of the cosmic machine revealed by science" (Quine 1969, p. 133), that is, the mechanistic constitution of the universe, however, we are able to transcend this inborn quality space and adopt a spacing of qualities which corresponds to nature's own spacing.[2]

These are perhaps Quine's first attempts to address a rather straightforward question suggested by his notion of empirical justification. The quality spaces determine the meaning of observation sentences, which in turn serve as checkpoints for scientific theorizing. Hence, the question arises: are these innate quality spaces good guidelines for scientific theorizing? The answer that Quine gives in *Natural Kinds* is: no, they are not. Therefore, he proposes that scientists have to be able to transcend this irrational spacing to properly engage in science. Put differently, according to the answer given in *Natural Kinds*, the success of science is made possible by our ability to rid ourselves of the animal sense of similarity and thus to move on from "unreason into science" (Quine 1969, p. 138), that is, ultimately, from a pre-scientific way of conceiving the world to a scientific one.

Note that if a community acquires this new spacing of qualities, this new spacing must exist along with the primitive one. Otherwise, infants could

[2]Quine's use of phrases such as *the cosmic machine revealed by science* (Quine 1969, p. 133) is certainly notable. I discuss below (Sect. 7.3.2), why the use of these metaphors does not imply that Quine defends anything close to contemporary scientific realism (for this contemporary notion, see below, Sect. 7.3.1). Still, these phrases explain why naturalistic realists like Kornblith (2002) are particularly fond of *Natural Kinds*.

not learn the language of their parents, as infants, being far from mastering theoretical science, have no clue about the non-primitive spacing.

To sum up, in *Natural Kinds*, Quine (1969) substantially revises the notion of quality space, or sense of similarity, when compared to *Word and Object*. He introduces the notion of theoretical similarity to explain why our science is successful despite the ineptitude of our innate sense of similarity for proper science. Furthermore, he does not employ the notion of stimulus meaning, which was central to his account of meaning and evidence in *Word and Object*. Instead, he often uses the vague notion of presentations when saying what this sense of similarity operates on. Probably, Quine, who is otherwise deeply committed to clarity, uses this vague term because it allows him to avoid answering a question which is unclear to him at the time of *Natural Kinds*: what are the inhabitants of innate quality spaces? They cannot accommodate physical objects because this would betray his entire naturalized epistemology; neither can they accommodate patterns of stimulations, as they differ too much among humans to serve Quine's purposes.

3.1.2 The Advent of Innate Perceptual Similarity Standards

In this section, I argue that the innovation that Quine was still missing in *Natural Kinds* is introduced with the notion of *perceptual* similarity standards in *The Roots of Reference*. In this work, Quine dedicates the entire first chapter to a thorough analysis of the concepts of perception and of perceptual similarity.

Quine's introduction of the notion of perceptual similarity marks a deliberate departure from the notion of similar stimulus meaning, which dominated in *Word and Object*. Consider the following passage, where Quine reflects on the kinds of similarity which can be applied to the episodes of a subject's life (an episode of a subject's life is roughly the entire material constitution of an individual human at a given point in time):

> Episodes in the subject's life ought, one feels, to admit of three sorts of similarity. There is receptual similarity, having to do only with sensory input. There is behavioral similarity, at the other extreme, which relates episodes according to the output of overt behavior at those times, regardless of causal factors. [...] Perceptual similarity, finally, should be somehow intermediate between receptual and behavioral similarity. It should be reflected in the behavioral output of the episode rather than just current input, but it should be reflected in only so much of the behavioral output as is *somehow distinctive to the current input* [emphasis RG]. (Quine 1974, p. 21)

According to this passage, receptual similarity is located on the level of stimulus meaning: it is a purely physiological measure of the sensory activity on a subject's nerve endings. Obviously, we can speak of similar stimulus meanings in the sense of receptual similarity: two stimulus meanings are more or less similar to each other depending on how similar the respective global stimuli are to each other—and the similarity of global stimuli is clearly a matter of input (in the sense of impacts of particles on sensory surfaces).

Behavioral similarity operates on a different level, to wit, on the level of the reaction of the subject to the stimulations. Two stimulations are judged more or less similar for a subject depending on the reaction of the subject—regardless what patterns irradiate his or her nerve endings. Quine (1974, p. 21) proposes an austerely physiological measure for behavioral similarity: "the total set of fibres of striped muscle that are contracted or released on one occasion and on another".

Quine then introduces his third kind of similarity, namely perceptual similarity. Perceptual similarity is intermediate between receptual and behavioral similarity insofar as it is (1) observable in the subject's behavior and (2) relates this behavior to aspects of receptual input. As I will detail in the following paragraphs, Quine tries to deal with the first requirement by conceiving of perceptual similarity as a second-order disposition to behavior. The second requirement is fulfilled by perceptual similarity's being a disposition to acquire dispositions to act in a certain way in response to certain patterns of sensory stimulations.

As detailed above (Sect. 2.1.3), Quine conceives of dispositions as a kind of scaffold for the progress of science: by means of dispositions, we can earmark effects that we suspect have a common mechanism as their cause, without yet knowing the respective mechanism. Thus, to conceive of perceptual similarity as a kind of disposition means that we earmark a class of phenomena that we suspect are the characteristic effects of a physical mechanism (or several such mechanisms), which we do not yet understand.

The characteristic effect of perceptual similarity, conceived as a second-order disposition, is that it causes and structures first-order dispositions. To return once again to the example of the parent, the child, and the parent's uttering "Red!", perceptual similarity enables the child to acquire the disposition to respond with "Red!" whenever it perceives a red surface. The child acquires this latter disposition because she is born with the second-order disposition to acquire first-order dispositions that closely resemble those of the parent.

Note that Quine no more holds that similarity standards operate on global stimuli of the subject's sensory surfaces, but rather on "brief stages or temporal segments of the perceiving subject's body" (Quine 1974, p. 16). This includes not only the irradiations of the subject's sensory surfaces, but also neural signals conveyed to his brain originating from his stomach, say, after a good meal. Hence, a given subject could be too sleepy to respond to a given stimulus in the way it has been conditioned to. According to Quine (1974, p. 22), this "ceteris paribus" effect is characteristic of dispositions. However, according to Quine, natural selection keeps such ceteris paribus effects under control in the case of perceptual similarity. After all, if we generally did not follow our "primitive inductions" (ibid.) acquired based on perceptual similarity (if we, for instance, do not run away from an approaching lion upon seeing it, but decide instead to give our stomach the desired rest), we would be in serious trouble.

Apart from keeping the *ceteris paribus* effect at bay, natural selection is also helpful in assuring that our innate sense of perceptual similarity runs "with the grain of nature" (and now it is obvious that we have moved from the genetic perspective to the justificatory perspective):

> After all, man's inductive expectations are reached by extrapolating along lines of perceptual similarity: experiences that begin similarly are expected to turn out in similar ways. Our innate standards of perceptual similarity show

> a gratifying tendency to run with the grain of nature. This concurrence is accountable, surely, to natural selection. Since good prediction has survival value, natural selection will have fostered perceptual similarity standards in us and in other animals that tend accordingly. (Quine 1974, p. 19)

After the negative assessment of our innate similarity standards in *Natural Kinds*—Quine wrote of the "brute irrationality of our sense of similarity" (Quine 1969, p. 125)—it is surprising how respectfully Quine treats these same standards in *The Roots of Reference.* He still believes that our innate sense of similarity changes over time, but it is obvious that he has abandoned the strict distinction between a subjective, animal sense which is innate, and a much more objective, theoretical sense which is required for scientific theorizing.[3] Furthermore, Quine holds that, since the innate standard of similarity is hereditary, "we can count on considerable social uniformity in perceptual similarity standards" (Quine 1974, p. 23). This means that it is rational to assume that human beings have very similar innate perceptual similarity standards.

It is remarkable how vaguely Quine expresses himself with regard to the epistemic capacity of our innate perceptual similarity standards. In the passage quoted, Quine writes that they "run with the grain of nature"—without ever specifying how to read this metaphor. Similarly, immediately before this passage, Quine writes that the standards have "a degree of objective validity" (Quine 1974, p. 19), again without explaining this claim. Given Quine's usual clarity and explicitness in his writing, this should raise eyebrows. The explanation for this vagueness that I favor is

[3] We still find these two positions in contemporary debates. The view of naturally selected abilities as irrational is basically the line of reasoning developed by Alvin Plantinga in his *Evolutionary Argument Against Naturalism.* In short, Plantinga (2002) argues that if our cognitive faculties are the product of evolution by natural selection, it is probable that naturalism is false, as evolution does not select for truth-conducive abilities, but only for abilities which foster survival. Since the latter are distinct from the former, evolutionary psychologists must conclude that, while their Darwinism may help them survive, it has nothing to do with truth. In contrast, Kornblith explicitly argues for a continuity between the kind of reasoning that fosters survival and the kind of reasoning that is at work in successful scientific theories (Kornblith 1993, pp. 61ff.; 2014, p. 132). Furthermore, Kornblith (2014, p. 126) makes clear that he commits himself to the claim that "natural selection is selecting for knowledge-acquiring capacities – that is, processes of belief acquisition that tend to produce truths". In contrast to both positions, Maddy (2017, p. 219) and Burge (2010, p. 308) hold that some mental ability's being the product of natural selection is neutral with regard to the ability's capacity to, as it were, deliver truths about the world.

that Quine is no scientific realist in the contemporary sense of the term; hence, he would not use a clearly realist terminology. Still, he thought that natural selection does give some epistemic validity to these evolved standards. Therefore, he uses a phrasing that gestures at such epistemic validity without committing him to anything like scientific realism (I discuss Quine's rejection of scientific realism below, Sect. 7.3.2).

To conclude this discussion, it is necessary to point out that, strictly speaking, it is not perceptual similarity, but an observation sentence's similarity basis which replaces the former notion of stimulus meaning. The similarity basis of an observation sentence is "the distinctive trait shared by the episodes appropriate to that observation sentence; the shared trait in which their perceptual similarity consists" (Quine 1974, p. 43). Quine here explains that the similarity basis of an observation sentence is the small part common to all the various sensory episodes in which it is correct to utter a given observation sentence. Quine calls this part the *shared trait* of these episodes. Note that the similarity involved in defining similarity bases is not receptual, but perceptual. Hence, the relevant episodes can be receptually very dissimilar, as long as they are perceptually similar. Accordingly, Quine uses clearly phenomenological language to illustrate his conception of trait: he writes that in the case of "Red!" the child has to learn that it is a matter of color, not of shape, of how big a patch counts as red, etc.[4]

Thus, unlike reception, which is "flagrantly physical", perception has "mentalistic overtones", but fortunately, it is "accessible to behavioral criteria" (Quine 1974, p. 4). In other words: even though perceptual similarity still operates on physically describable events in a subject's body, there is, as it were, more to perception than meets the receptors. Quine (1974, pp. 17–18) therefore proposes a behavioral test to delineate a subject's sense of perceptual similarity. In this test (which is reminiscent of the one involving an animal, a lever, and a pellet of food discussed above, Sect. 2.1.3), a subject's perceptual similarity standards can be measured as follows. We condition the subject to respond in a certain way to a certain stimulus. Furthermore, we condition the subject *not* to respond

[4]This has led Føllesdal (2014, p. 276) to claim that Quine's position is surprisingly similar to Husserl's.

thus to another stimulus. Then, we can test whether a third stimulus is perceptually more similar to the first than to the second for the subject by stimulating the subject in the third way and observing whether the subject does or does not react in the manner in which it has been conditioned to react to the first stimulus.

Still, Quine's notion of perceptual similarity is not in the same sense physical as his notion of stimulus meaning. While, in the case of stimulus meaning, Quine could actually say what the physical structures and mechanisms consist in, perceptual similarity is rather an "earmark" (Quine 1974, p. 11): it refers to a class of phenomena that we suspect are the effects of a physical mechanism, or of several such physical mechanisms.

Despite this introduction and detailed discussion of perceptual similarity, Quine expresses a residual unease about the privacy of stimulus meaning in *The Roots of Reference.* Quine holds that a non-verbal common ground between the parent and the infant is needed to get the process of first-language acquisition started: "somewhere there have to be nonverbal reference points, nonverbal circumstances that can be intersubjectively appreciated and associated with the appropriate utterance on the spot" (Quine 1974, p. 37). One page later, he hits upon the problem that observations are "sensory, evidently, and thus subjective"; this leads him, another page later, to propose to abandon talk of observation altogether by only speaking of observation sentences, since they represent what competent speakers actually agree on (Quine 1974, pp. 38–39). Quine (1974, p. 39) then gives the following definition of observation sentences: "A sentence is observational insofar as its truth value, on any occasion, would be agreed to by just about any member of the speech community witnessing the occasion." According to this definition, all that is needed for an observation sentence is a unanimous agreement on its truth value in any situation.

At the end of the section, however, he admits that this does not solve the problem:

> The definition does raise a subtler problem, however – a problem that was already noted in another connection in §6. The definition speaks of joint witnessing. In a more precise statement, it would speak of witnesses subject to receptually similar impingements; and thus it would raise again the homology question that we noted at the end of §6.

> Still, the definition is as sharp as the notion of witness and linguistic community on which it depends. It is good [sic] as behavioral concepts go. (Quine 1974, p. 41)

As Quine rightly points out in this passage, joint witnessing, a notion which he needs to define observation sentences, requires that witnesses can be similarly stimulated and hence provokes again the homology problem.[5] In short, at the end of the discussion in *The Roots of Reference*, Quine admits that he has not solved the basic problem caused by the fact that stimulus meaning is private (his gesture at the intrinsic limitations of behavioral concepts is inconclusive; Quine himself was not satisfied with this response to the issue and continued to belabor it).

3.1.3 An Emphatic Interlude

18 years after the publication of *The Roots of Reference*, Quine's *Pursuit of Truth* (which is here quoted in its second edition) takes up again the issue of shared stimulus meanings. In the course of discussing radical translation, Quine remarks that the linguist and the native speaker share a common situation on which the linguist can base her translation manual. Radical translation is Quine's term for the task of a linguist to develop a translation manual for a language that is entirely unknown to her and unrelated to any language that she already knows. Furthermore, Quine (1992, pp. 39–40) notes that "in earlier writings" he used to cash out this notion of common situation by the use of "same stimulations", and

[5]See also Quine (2016 [1995], pp. 29ff.), where Quine emphasizes publicly observable behavior at the cost of intersubjectively shared stimuli—and then gives an account of how preestablished harmony of innate perceptual similarity standards is needed to explain first-language acquisition based on such publicly observable behavior. This nicely agrees with my position, according to which the preestablished harmony is needed to, as it were, compensate for the privacy of stimulus meanings (see below, Sect. 3.1.4). I take the following paragraph from said article (Quine 2016 [1995], pp. 30f.) to express exactly this problem caused by the privacy of stimulus meanings:

> In Pursuit of Truth that appeal [to intersubjective similarity of stimulus meanings, RG] is out of order, so the causal question recurs: why, after the mother has got the child to affirm 'Milk' once in an appropriate situation, does the child's usage continue to agree with the mother's? The answer can lie no longer in intersubjective similarity of stimulus meaning.

that this notion depends on "sameness of stimulation of both parties, the linguist and the informant". Finally, in a sequence reminiscent of *The Roots of Reference*, he first discards the idea of sameness of stimulations, since the two parties do not share any receptors. Then he abandons the idea that the two parties are even similarly stimulated, since this would still presuppose approximate homology, but, as Quine puts it: "such anatomical minutiae ought not to matter here". That is, such anatomical details should be irrelevant for his project of naturalized epistemology.

Subsequently, Quine (1992, p. 42) declares that likeness of stimulation is not required to help the linguist with the translation of the first observation sentences. This abandonment of sameness of stimulus meaning blurs the "distinctive factuality" (Quine 1992, p. 43) of observation sentences, that is, it eliminates a strict distinction of observation sentences from other sentences in physicalistically acceptable terms. However, Quine submits, it leaves the intersubjective character of observation sentences intact:

> Unlike Davidson, I still locate the stimulations at the subject's surface, and private stimulus meanings with them. But they may be as idiosyncratic, for all I care, as the subject's internal wiring itself. What floats in the open air is our common language, which each of us is free to internalize in his peculiar neural way. Language is where intersubjectivity sets in. Communication is well named. (Quine 1992, p. 44)

In this passage, Quine reaffirms, in explicit opposition to Davidson, that he conceives of the stimulus and the stimulus meanings as located on the subject's surface, that is, as patterns of stimulations of the subject's nerve endings. Furthermore, in contrast to the picture of *Word and Object*, he acknowledges that these stimulus meanings are private, not intersubjectively shared. What ensures intersubjectivity and hence communication is language itself. In this sense, it floats not only in the open air, but even in a void: there is nothing else that is equally intersubjective that could underpin the intersubjectivity of language.

Coming from *The Roots of Reference* (and, as will become evident in the following, also from the approach adopted in *From Stimulus to Science*), one would expect now an introduction of perceptual similarity. Instead, immediately after declaring that stimulus meaning is private, Quine (1992,

p. 42) claims that *empathy* plays a central role in language learning: "Empathy dominates the learning of language, both by child and by field linguist. In the child's case it is the mother's empathy."

Unsurprisingly, for Quine (1992, pp. 42–43), 'empathy' means a straightforward, behavioristically assessable ability. It is essentially the ability to accurately model how a given situation appears to one's counterpart.

How can empathy, so understood, support Quine's account of first-language acquisition? Once again, imagine the situation in which the parent tries to teach the observation sentence "Red!" to the child. In this situation, Quine (1992, p. 42) maintains that the parent uses her empathy to imagine how the scene presents itself from the point of view of the infant: "[T]he parent assesses the appropriateness of the child's observation sentence by noting the child's orientation and how the scene would look from there". Further, in the paragraph on observation sentences, Quine requires that an observation sentence be "directly and firmly" associated with its stimulus meaning (which may vary from speaker to speaker, see Quine 1992, pp. 3–4). Thus, the correct use of "Red!" is still a good place to start first-language acquisition because it is very strongly associated to (language-wise presuppositionless) patterns of sensory stimulation. When being conditioned to respond with "Red!" only to appropriate stimuli, the infant can count on the parent's empathy: whenever he tests a new patch of color and utters "Red!", the parent puts herself in the infant's place and determines whether she would have made the same statement.

However, the mere intersubjectivity of language, even when supplemented with empathy, cannot explain the process of first-language acquisition. Between the pre-linguistic infant and his parent, there is no intersubjectively shared language. This makes clear that Quine's conception of empathy cannot replace the role of socially uniform perceptual similarity standards in first-language acquisition. At most, empathy can help to spell out a process that presupposes these standards. Only if such standards are in place can the parent's empathy reasonably come into play in the process of first-language acquisition. Unless both parent and child find the same situations similar, the parent's putting herself in the child's situation will not help.[6] This, finally, leads back to the problem of joint witnessing that Quine himself has pointed out (see the end of the preceding section).

With regard to the way in which scientific theories are empirically justified, the same problem has different implications. It seems difficult to see how abandoning any recourse to sameness of stimulations could make the scientist's agreement on observation sentences more than a merely verbal agreement. Compare the following passage, where Quine sketches the main question of naturalized epistemology:

> From impacts on our sensory surfaces, we in our collective and cumulative creativity down the generations have projected our systematic theory of the external world. Our system is proving successful in predicting subsequent sensory input. How have we done it? (Quine 1992, p. 1)

The use of the pronoun 'our' in this passage is ambiguous. On the one hand, when writing about the systematic theory, it expresses, as it were, a common possession, something which belongs and which is accessible to an entire community. The theory of the external world is 'our' theory in the sense that there is one single theory which belongs to and is the product of an entire community. This use is similar to the use in the phrase 'our health care system': we, as a community, finance and possess one single system which is here for all of us when we need it. On the other hand, the sensory surfaces belong to us in a different, individual sense. Each of us has their own sensory surface. This use of 'our' is the same as in 'our health'. Strictly speaking, we are not talking about our health, but about the individual health of each member of the class to which 'our' refers. Now the important point is that the sensory input which is predicted by the theory is always 'our' input in the second sense, which is *relative to a given subject.* Conversely, after using the plural in the first sentence ("sensory surfaces"), Quine should have used it also for 'sensory input'. The only term that is entitled to a singular is 'theory'.

---

[6] Remember how Quine (1960, p. 83) puts it: "If the child is to be amenable to such training, however, what he must have is a prior tendency to weight qualitative differences unequally. He must, so to speak, sense more resemblances between some stimulations than between others. Otherwise a dozen reinforcements of his response 'Red', on occasions where red things were presented, would no more encourage the same response to a thirteenth red thing than to a blue one".

In short, the verdict of *The Roots of Reference* still stands: in the absence of nonverbal reference points, for instance intersubjectively shared stimulus meanings, Quine cannot explain how language is acquired, nor can he legitimately claim that the agreement of scientists is more than a merely verbal agreement. In the next section, I show that in his last monograph, *From Stimulus to Science*, Quine is acutely aware of both issues, and I delineate the monograph's bold proposal to resolve them.

3.1.4 Finally: A Preestablished Harmony of Similarity Standards

Only three years after the publication of *Pursuit of Truth* in its second edition, in *From Stimulus to Science*, Quine presents an account of the infant's first steps into language which is in many ways closer to *The Roots of Reference* than to *Pursuit of Truth*: empathy does not appear until the last chapter, while innate standards of perceptual similarity take center stage. Furthermore, Quine assigns evolutionary considerations a central role in his attempt to explain how communication is possible. For 'communication' to be possible requires that the recipient of a certain message can in principle understand what the sender of the message means by the message. With intersubjectively shared stimulus meanings, Quine could account for this requirement in the case of observation sentences. Without them, Quine runs the danger of being committed to the view that speakers are necessarily talking past each other even in the case of observation sentences.

Quine introduces *receptual* and *perceptual similarity* in close analogy to his introduction of these notions in *The Roots of Reference*. The receptual similarity between two global stimuli simply measures the extent to which, in these two global stimuli, the same receptors are stimulated at the same time. Perceptual similarity, Quine (1995, p. 17) maintains, is a matter of "effect on the subject, a question of reaction". Thus, the perceptual similarity between two global stimuli cannot be 'read off' from the stimulations of the subject's nerve endings. Rather, it has to be assessed by means of the now well-known behavioral experiment (compare above, Sects. 2.1.3 and 3.1.2). Briefly, the experiment proceeds as follows. Stimulate a subject in a certain way and reward its reaction. Then, stimulate it in a receptually different way and penalize the same reaction. Finally, stimulate it in a way which is receptually somewhere between the two earlier stimulations and see whether or not the subject reacts or not. If it does not, the third stimulation is closer to the second, etc.

The third concept to be introduced is *salience*. The salient receptors in a given global stimulus are those that matter. To return to my favorite example, the infant has got the trick with the observation sentence "Red!" if he, as it were, infers that it is exactly the retina's stimulation by those light waves which are reflected from the red surface that are salient at the moment. The salient parts of two perceptually similar global stimuli for a given subject can be obtained by subtracting these parts of the stimuli which are receptually dissimilar (Quine 1995, p. 18).

Because standards of perceptual similarity are required for first-language acquisition, indeed, for all learning, we are born with a provisional standard of perceptual similarity. After enlisting natural selection for a first time to ensure that our innate standards "mesh pretty well with natural trends", Quine (1995, p. 20) rephrases the puzzle of shared meaning and also sketches the requirements for a solution[7]:

> There is a puzzle here. Global stimuli are private: each is a temporally ordered set of some one individual's receptors. Their perceptual similarity, in part innate and in part molded by experience, is private as well. Whence then this coordination of behavior across the tribe? It requires that if two individuals jointly witness one scene, and subsequently jointly witness another scene, and the one witness's global stimulations on the two occasions qualify for him as perceptually similar, usually the other witness's stimulations will also so qualify for the other witness. (Quine 1995, p. 20)

In this passage, Quine first reaffirms the presuppositions of the puzzle. In contrast to the account given in *Word and Object*, global stimuli are private, and so is perceptual similarity. The problem emerges if these private entities should function as the building blocks of something decidedly

[7]The vague expression "mesh pretty well with natural trends" is another Quinean attempt to say that evolution ensures that these standards have a certain epistemic validity, without committing himself to anything like scientific realism. Compare above, Sect. 3.1.2 and below, Sect. 7.3.2.

public, namely of language. If we experience situations differently, and if we find different experiences similar, it seems impossible to explain how we have managed to develop a common language, and it seems deeply mysterious that we should even agree on what to say regarding a specific situation. Furthermore, Quine gives a very succinct wording of what is required to dispense with this mystery. If two speakers of a language witness two scenes, they must unanimously perceive the respective global stimuli as perceptually similar or not.

In accordance with his overall naturalized epistemology, Quine makes global stimuli, and not situations, the objects of comparisons. The requirement, however, irreducibly involves these shared situations: the two speakers must agree on their verdict regarding global stimuli of two situations or scenes. Thereby, the main challenge is that these two situations give rise to four different global stimuli—one global stimulus for each speaker in each situation. Therefore, in each of the two situations, speaker *A* must consider two global stimuli as similar (or dissimilar) which differ from the stimuli that speaker *B* must judge as similar (or dissimilar). The challenge therefore is not simply to explain why speakers find *the same* situations perceptually similar, but rather why they find *the incongruent* respective global stimuli that arise from these situations perceptually similar. Strictly speaking, the scenes for the two speakers are never the same, since they always experience different global stimuli.

A first step to solving this puzzle has already been taken in *The Roots of Reference*: Quine proposes to focus on *perception*, as opposed to *reception*. Quine's conception of perception allows for perceptual similarity in spite of receptual dissimilarity. In particular, there is no principled reason why two subjects should not find receptually dissimilar situations perceptually similar. The crucial question therefore is: why should these innate perceptual similarity standards be so constituted that the speakers will, in spite of receptually dissimilar global stimuli, perceive the same situations as similar? Quine tackles this question in the following passage:

> So we see a preestablished harmony of perceptual similarity standards. If two scenes trigger perceptually similar global stimuli in one witness, they are apt to do likewise in another.

> This public harmony of private standards of perceptual similarity is accounted for by natural selection. The individual's initial standards of perceptual similarity are inculcated, we saw, by natural selection, and so, thanks to shared ancestry and shared environment, will tend to harmonize across the tribe. The changes in standards subsequent to birth will also tend to harmonize, because of the shared society and environment. (Quine 1995, p. 21)

According to this passage, there is a preestablished harmony of perceptual similarity standards among human beings. This preestablished harmony ensures that, if two scenes cause perceptually similar global stimuli in one human, they typically also cause perceptually similar global stimuli in any other human.

Furthermore, Quine claims that this preestablished harmony can be explained by natural selection, as far as the innate perceptual similarity standards are concerned. These standards stand in a preestablished harmony with our fellow human beings' because they are genetic traits which are inherited from our common ancestors, and which have developed in a shared environment.

Finally, these parts of the standards that are not innate are uniform among a speech community because of "shared society and environment". This means that they are uniform because we live in the same society and in the same environment.

Quine (1995, p. 22) then introduces observation sentences in accordance with how he has introduced them in *The Roots of Reference*: (1) they are a kind of occasion sentences. (2) All members of a language community agree on the verdict on such a sentence because they report intersubjectively observable situations. (3) For every subject, these observation sentences are tied to "a range of perceptually fairly similar global stimuli", that is, for every subject, each observation sentence is directly associated with a range of perceptually similar global stimulations of her nerve endings. Two global stimuli can be overwhelmingly, almost entirely dissimilar in terms of reception, but still very similar in terms of perception if the salient part of the global stimulus is very small (as it often is). Finally, Quine (1995, p. 22) argues that the preestablished harmony of perceptual similarity standards ensures that the privacy of the global stimuli and of perceptual similarity standards do not inhibit the community-wide agreement on any observation sentence.

According to this account, first-language acquisition works *in spite, and not because* of the patterns of sensory stimulations that affect both the infant and the parent. As shown above, for Quine, the infant and the parent do not witness the same situation; rather, they are experiencing different global stimuli in the same situation (the sameness of the situation being irrelevant both for the genetic and the justificatory aspect of Quine's investigation). This means that the Quinean subject, while not being causally entirely windowless, like the Leibnizian monad, is still experiencing sensory stimulation *as something that separates her from, rather than connects her to,* her fellow humans.

Quine cherishes the preestablished harmony of innate perceptual similarity standards because he thinks they allow him to avoid attributing any perception as of physical objects to the infant to explain her acquisition of her first observation sentences. By the same token, he thinks that this preestablished harmony allows him to explain how real communication is possible in spite of the privacy of stimulus meanings. Distal objects would be natural candidates as intersubjective checkpoints for both language learning and scientific theorizing; however, as I have shown above (Sect. 2.2.2), assigning such a role to distal objects contradicts the basic set-up of Quine's naturalized epistemology.

3.2 Summary: Quine's Naturalized Epistemology

The infant who is learning her first language and the scientist who is testing her theories have the same evidence available for their respective tasks, namely observation sentences and the stimulus meanings associated with them. Quine studies first-language acquisition guided by two basic presuppositions, namely eschatological physicalism, which suggests his behaviorism and compels him to reject final causes, and his source-empiricism.

Quine's *Word and Object* is his first serious attempt to carry out naturalized epistemology. There, Quine develops an elegant and innovative physicalistic notion of linguistic meaning and, by the same token, of empirical data. An observation sentence, taken holophrastically, is directly associated with its stimulus meaning, consisting of specific patterns of sensory stimulations. This stimulus meaning of holophrastic observation sentences is intersubjectively (almost) invariant. Furthermore, in its holophrastic reading, an observation sentence is not structured in any (proto-)logical way, and it is not ontologically committed. This means that it does not presuppose any linguistic knowledge. Thus, holophrastic observation sentences can serve as the gateway to language for the child and as theory-neutral touchstones for scientific theorizing.

Quine explains the process by which the infant learns her first observation sentences in austerely behavioristic terms: the infant is conditioned to respond by uttering a given observation sentence when being stimulated in the appropriate way. For this conditioning process to be possible, the infant needs an innate quality space that is biased such as to privilege these aspects of stimulations that matter for language learning.

Around 1965, however, Quine realizes that it is wrong to suppose that stimulus meaning could be intersubjectively shared, since humans' nerve endings are non-homologous. This means that his account of *Word and Object* cannot explain the intersubjectivity of language. In *Natural Kinds*, Quine cannot present a replacement for intersubjectively invariant stimulus meaning. This explains his noncommittal use of 'presentations', which refers to the inhabitants of quality spaces. Furthermore, in *Natural Kinds*, the epistemic significance of these quality spaces comes into focus for the first time. Quine entertains the idea that our innate quality spaces are brute and irrational, which means that they have to be overcome to engage in serious scientific theorizing.

The Roots of Reference contains Quine's introduction and most elaborate discussion of innate perceptual similarity standards. Perceptual similarity standards are second-order dispositions, which shape first-order dispositions to behavior. The relevant behavior here is uttering an observation sentence in the correct situation. For a given subject, whether or not it is appropriate to utter a given observation sentence in a certain situation is decided by the observation sentence's similarity basis. This similarity basis is the distinctive sensory trait shared by all sensory episodes in which it is correct to utter the respective observation sentence. The infant is able to identify this trait thanks to her innate standards of perceptual similarity.[8] A notable difference between *Natural Kinds* and *The Roots of Reference* is that Quine maintains only in the latter work that the innate similarity standards run with the grain of nature (I have suggested that the vagueness of this phrase is deliberate). These elaborate and subtle conceptions of innate perceptual similarity standards and similarity basis notwithstanding, Quine admits that there must be nonverbal reference points somewhere, which he is, at the end of the day, still unable to provide.

In *Pursuit of Truth*, there is no mention of perceptual similarity standards; instead, Quine develops the notion of empathy to explain first-language acquisition in the absence of intersubjectively invariant stimulus meanings. I have suggested that empathy cannot replace these perceptual similarity standards but rather presupposes them and, hence, again provokes the question of nonverbal reference points.

Finally, in *From Stimulus to Science*, Quine reinstalls perceptual similarity to the central role it played in *The Roots of Reference.* Quine suggests that natural selection ensures a preestablished harmony of different human beings' innate perceptual similarity standards, which enables an infant to acquire her first language. Natural selection also guarantees that these standards "mesh pretty well with natural trends" (Quine 1995, p. 20).

I have argued that a preestablished harmony of perceptual similarity standards can indeed replace the intersubjectively shared stimulus meaning of *Word and Object*: the preestablished harmony guarantees that whenever a human perceives two global stimuli of two situations as similar, another human present in both situations will also perceive his respective stimuli (which are dissimilar to the ones of the first human) as similar. This can explain the existence of communication insofar as it provides a common ground between the two speakers that is not merely verbal: both speakers, and indeed all competent speakers, assent and dissent to a given observation sentence in the same two classes of situations. This makes scientists' agreements on observation sentences more than a merely verbal

[8]This summary of the account from *The Roots of Reference* is mentalistic. This mentalistic vocabulary is also used by Quine (1974, p. 43); of course, he would argue that he is not committed to the mentalistic version of the story, but only to the safely behavioristic one. However, I will suggest below, Sect. 3.3.2, that he is not able to completely avoid intentional vocabulary.

agreement, and it provides the infant with the indispensable non-verbal reference points to acquire her first language.

However, it does so by diminishing the explanatory role of patterns of sensory stimulations and by greatly increasing the role of innate perceptual similarity standards. Indeed, on the account given in *From Stimulus to Science*, the fact that any two speakers of a language unanimously agree on their verdict regarding the observation sentence "Red!" in any given situation is not explained with recourse to intersubjectively invariant stimulus meaning. Rather, natural selection guarantees that the innate perceptual similarity standards of the two speakers stand in a preestablished harmony. Quine maintains that this harmony explains their agreement *in spite of* the respective stimulus meanings (which are necessarily dissimilar). With regard to the theoretical background sketched in Sect. 2.1, this means that Quine has made quite a development from *Word and Object* to *From Stimulus to Science*: both empiricism and physicalism are less directly implemented in his mature proposal (indeed, it is not clear whether this proposal still deserves to be called empiricist, given the central role played by innate cognitive structures).

Let me phrase the same point differently. It is an innate structure, namely perceptual similarity standards, which explains the intersubjectivity and hence the very possibility of communication. This intersubjectivity presupposes that speakers of a common language typically mean the same thing when they are using observation sentences—that they are not constantly talking past each other because they do not associate the same meanings with these observation sentences. This is central in particular if the agreement of scientists on observation sentences is to be more than merely verbal agreement, and if the infant is to have some non-verbal, intersubjectively shared reference points that allow her to acquire her first language. Both requirements can be satisfied by invoking Quine's preestablished harmony of similarity standards. Thanks to them, any two speakers perceive the same situations as similar or not, and hence are prepared to assent to a given observation sentence in the same situations. The nonverbal reference points that satisfy both requirements are therefore given by the shared receptual traits of stimulations that are perceived as similar.

This also means that Quine has already most of the material for this solution available in *The Roots of Reference*. It is not until *From Stimulus to Science*, however, that Quine is prepared to take the plunge and give such epistemic and genetic weight to these innate perceptual similarity standards.

To conclude, let me conceive this final attempt of Quine to come to terms with the intersubjectivity of communication and with the success of science in the more general framework of naturalized epistemology. First, how does Quine explain the emergence of successful science? Regarding the part of Quine's answer that applies to the very first steps of the emergence of science, the answer is as follows. Humans are born with innate perceptual similarity standards. These standards stand in a preestablished harmony with their fellow humans' standards and hence make communication, also scientific communication, possible, in spite of private stimuli: thanks to these standards, any two humans will perceive the same situations as similar or dissimilar respectively. Furthermore, the science that builds on these standards is successful because the standards have been molded by evolution such that they, to use Quine's words, mesh pretty well with natural trends.

Second, which conclusions can be drawn with regard to the justification of scientific theories, or—to put it more Quinean—with regard to the relation of theory to evidence?

I will examine this question in-depth when discussing Quine's defense of empiricist justificatory monism (see below, Sect. 6.1). For now, note that the evidence for scientific theories, according to the account reconstructed in the present chapter, is very limited. Quine (1992, p. 8) maintains that the evidence consists in holophrastic observation sentences that are directly associated with a given, private pattern of stimulation. Via such observation sentences, as Quine (1992, p. 36) puts it metaphorically, the world gives its verdict on our theories.

Hence, even if the stimulus meaning of a given observation sentence is held together by perceptual, and not receptual similarity, the meanings of observation sentences, and hence the checkpoints for scientific theorizing, are still given by patterns of stimulations: it is, for any given observation sentence and relative to a given subject, the shared receptual trait that is common to all situations in which the subject would assent to the respective observation sentence.

3.3 Two Internal Problems for Quine's Eschatological Physicalism

In this section, I first critically assess Quine's invocations of natural selection (Sect. 3.3.1). Then, I argue that his account of how the infant acquires her first observation sentences makes tacit references to communicative purposes. Furthermore, I suggest that, once purged of these communicative purposes, his account becomes highly implausible (Sect. 3.3.2).

3.3.1 Quine's Invocations of Natural Selection Are Probably Unscientific

When affirming that natural selection guarantees the preestablished harmony of innate perceptual similarity standards, Quine enlists natural selection to solve a very specific problem in his elaborate account of communication. Regarding the solution to this problem, it is crucial to distinguish between some feature's being plausibly the result of a process of random variation and natural selection (because the feature is *prima facie* conducive to fitness) and there being a fully-fledged, evolutionary-biological account of *how* that same feature has resulted from this process.

It would certainly be unfair to ask Quine for such a fully-fledged account of the evolution of the preestablished harmony of perceptual similarity standards. After all, Quine (1974, p. 3) explicitly maintains that his project "may be pursued at one or more removes from the laboratory, one or another level of speculativity." In this vein, he could respond that his invocations of natural selection do not have to satisfy fully-fledged scientific standards, since he operates on the more speculative level, where he sketches a new research perspective that may then be fleshed out by trained specialists.

Hence, the interesting question is: is it plausible that it is possible for biologists to provide such a fully-fledged evolutionary-scientific account of the phenomenon in question? The phenomenon in question is the preestablished harmony of innate perceptual similarity standards, which is the cornerstone of Quine's account how genuine communication is possible at least on the level of holophrastic observation sentences.

In the remainder of this section, I raise doubts regarding this possibility. My entire discussion is heavily indebted to Richardson (2007), who insists that some subject matters are well-suited for evolutionary studies, while others are not:

> Fruit flies are a well-adapted subject for evolutionary studies. Likewise, as I illustrated in the previous chapter, the morphology of *Gammarus minus* is one that is amenable to evolutionary analysis. But psychology, in contrast, is poorly adapted as an evolutionary science. (Richardson 2007, p. 148)

Richardson here essentially claims that the methods that proved successful in the case of fruit flies and in the case of the morphology of *Gammarus minus* (a species of amphipod crustacean) are likely to be not equally successful when applied to psychology. The reason for this is that it is much more difficult to examine human psychology in a way that satisfies core methodological principles of evolutionary biology than it is to examine the morphology of *Gammarus minus* in such a way. As will become evident, this has much to do with the amount of evidence and information that is available about the morphology of *Gammarus minus*, but not with regard to human psychology.

In the following, I introduce these elements, or pieces of information, that an ideal explanation in evolutionary biology requires, following Richardson (2007, pp. 99–104; he himself refers to Brandon 1990, chap. 5). Richardson admits that an explanation by natural selection does not have to detail all of these pieces to pass muster; but they do give an idea of the "sort of criteria relevant to deciding whether an explanation passes muster" (Richardson 2007, p. 99). There are five such elements.

Selection. First, Richardson requires that "There must be evidence that selection has in fact occurred". This in turn requires that we have information regarding the variation of the feature in question by the time the natural selection is postulated to have occurred, together with information about the differences in terms of survival and reproductive success that these different features have bestowed upon their bearers.

Ecological Factors. Next, Richardson requires that we can explain said selection with recourse to ecological factors, that is, we need to be able to specify which properties of the social or physical environment caused a certain trait to be conducive to its bearer's reproductive and survival success.

Heritability. For evolution through natural selection to be possible, the features that increase the reproductive fitness of their bearers must be heritable. This means, according to Richardson (ibid.), that phenotypical features should be correlated between parent and offspring to an extent that is clearly greater than mere chance. This requires that if the parent has a specific selective advantage, it should be more likely than not that the parent passes this advantage on to the offspring.

Population Structure. The gist of this factor is that if there are different populations living in different habitats that furnish different biological features with a selective advantage, then a gene flow between these populations is likely to cancel out any selection due to the several different habitats. For instance, assume that there are two populations of fruit flies, one living by a pond, the other living in a swamp. The populations are sufficiently close to allow for frequent mating between members of the two populations. Furthermore, assume that *large* fruit flies living in the swamp have more offspring than small ones living in the swamp, while the converse holds for fruit flies living by the pond. In such a situation, the two selective pressures are likely to cancel each other out: there will be no significant increase or decrease in size of the fruit flies in either population across generations because of inter-population mating.

In contrast, a completely isolated population in a uniform environment that exercises a consistent selective pressure on the population is more hospitable to evolution through natural selection. For instance, assume that there is just the one population living in the swamp, where larger flies have significantly more offspring. Then, there may occur a selection and an increase in size of individual flies across generations.

Trait Polarity. Richardson uses this condition to emphasize that there is a difference between a feature's being *adaptive* and a feature's being *an adaptation.* In particular, there can be adaptive features that are not adaptations. This is the case if a certain biological feature that is very useful, perhaps even indispensable for the life-form of a given species, is present in ancestral species already. Richardson (ibid.) mentions the skull sutures of mammal infants. Without them, mammalian birth would probably be impossible, so this feature is very adaptive. However, since it is present in birds and reptiles, it cannot possibly be an adaptation to said mammalian birth: the skull sutures existed well before mammalian birth evolved.

To illustrate how these elements are included in state-of-the-art explanations by natural selection, Richardson (2007, pp. 120–124) presents the case of *Gammarus minus.* This species of amphipod crustacean has specific adaptations for its habitat, dark caves in West Virginia. Individuals belonging to this species differ from their non-cave-inhabiting ancestors mainly in being larger, having drastically reduced eyes, and having larger antennae. The question to be answered in the following is whether these differences can be explained with recourse to natural selection. Regarding *selection*, it is easy to verify that there is variation of the trait: one can simply measure body and antenna size and count the eye facets of individuals living in the caves. Then, one can compare these numbers to the ones of their relatives that do not live in caves but rather near springs.

Furthermore, in the habitat of *Gammarus minus*, which is completely dark, larger antenna are obviously conducive to survival. More precisely, concerning the *ecological factors* that drive the selection, Richardson (2007, pp. 122–123) mentions that field studies in caves show that "individuals with smaller eyes, larger antennae, and larger body size mate more frequently and produce more offspring than those in the same population with larger eyes and smaller antennae and body size." Regarding *heritability*, Richardson (2007, p. 123) introduces a laboratory study that indicates that offspring of cave-dwelling Gammarus have smaller eyes than offspring of spring-dwelling Gammarus, with a correspondence that roughly matches the heritability of body height in humans. Concerning *population structure*, it has been verified that the different populations of *Gammarus minus* are very isolated and exposed to a very homogeneously light-less environment (Richardson 2007, pp. 123–124). Trait polarity, finally, is very easily established, as the non-cave-dwelling *Gammarus* have substantially bigger eyes, but shorter bodies and antenna.

Now, let us consider Quine's explanation by natural selection of the preestablished harmony of humans' innate perceptual similarity standards. Of the five pieces of information that Richardson requires from an ideal explanation by natural selection, Quine only touches upon two: selection and ecological factors. Concerning selection, he does not specify the variation of similarity standards between individuals, nor does he specify the time where the selection is supposed to have occurred. In terms of the ecological factors that have conferred a survival advantage on their bearers, Quine does not get more specific than "the grain of nature" or "natural trends". There is no information on heritability, population structure, or on trait polarity. In short, it is obvious that Quine's invocations of natural selection fall short of the standards employed by the science of evolutionary biology.

However, as mentioned above, the interesting question is not whether Quine's own invocations fulfill these standards, but rather whether it is possible that evolutionary biologists develop an account along Quine's lines that fulfills these standards. There are good reasons to assume that this is not possible. Consider the five kinds of information that Richardson lists for an ideal explanation of a human trait by natural selection. Take as the trait in question the preestablished harmony of human perceptual similarity standards, which explains the possibility of communication. Skulls and bones cannot give us any information regarding the variation of innate perceptual similarity standards, since they are *behavioral* not *anatomical* traits. Similarly, we have no clue as to the ecological factors that shaped these standards, or about the relevant population structure and the trait polarity. More importantly, it seems very unlikely that we will ever have such information available.

Hence, while it is not impossible that, at some point, we will have all the information necessary for a proper evolutionary explanation to cash out Quine's speculations, or, alternatively, that the methodological standards for such an explanation (perhaps due to vastly more reliable computer simulations) change drastically, it is doubtful whether there will be such an explanation. This means that it is doubtful whether Quine is able to explain why there is such a thing as successful communication, even on the level of holophrastic observation sentences. Not because it is demonstrably wrong that natural selection can ensure that there is a preestablished harmony of perceptual similarity standards; rather, because it is almost demonstrably undemonstrable that natural selection can guarantee such a preestablished harmony.

3.3.2 Quine's Infant Needs to Understand Final Causation

I have already delineated above (Sect. 2.1.3) how Quine uses his behaviorism in psychology to cash out this eschatological physicalism. Behavioristic psychology, as Quine understands it, operates purely within the realm of stimulus and response patterns, and hence of efficient causes. As we have seen, on Quine's view, learning an observation sentence does not consist in understanding any mentalistic meaning, but merely in being subjected to a certain conditioning that then causes the infant to behave in a way that competent users judge as correct.

In this section, I argue that Quine's account of first-language acquisition tacitly employs purposes, namely communicative purposes. Furthermore, I argue that his account is implausible if it is purged of such communicative purposes. To do so, I focus on Quine's notion of pleasure through approval.

Quine's most detailed discussion of the role of pleasure, or reward, in the context of first-language acquisition is to be found in *The Roots of Reference.* The following passage is central:

> [1] Thus consider the learning of the word "red". Suppose the child happens to utter the word in the course of the random babbling that is standard procedure in small children, and suppose a red ball happens to be conspicuously present at the time. The parent rewards the child, perhaps only by somehow manifesting approval. [2] Thus in a certain brief minute in the history of overall impingements on the child's sensory surfaces there were these features among others: there were light rays in the red frequencies, there were sound waves in the air and in the child's headbones caused by the child's own utterance of the word 'red', there were the impacts on the proprioceptors of the child's tongue and larynx occasioned by that utterance, and there were the impacts, whatever they were, that made the episode pleasant.

> [3] On a later occasion a red shawl is conspicuously present. Its color makes for a degree of perceptual similarity between the pleasant earlier episode and the present, thus enlivening the trace of that episode. The child contorts his speech muscles so as to add what more he can to the similarity: he again says 'red', and we may hope that the similarity is yet further enhanced by a recurrence of the reward. (Quine 1974, p. 29)

To facilitate discussion, I have divided up this passage into three parts, marked by '[1]', '[2]', and '[3]'. [1] gives a condensed version of the passage from *Word and Object* (quoted at the beginning of Sect. 2.2.1, p. 43, see Quine 1960, pp. 80–81). The child unconsciously babbles something that sounds like "Red!"—and by mere chance there is a red thing present in a conspicuous way. The parent, who also happens to be present, witnesses this utterance of the infant and rewards it, "perhaps only" by "manifesting approval". In [2], Quine gives his physicalist-behavioristic reading of the scene sketched in [1]. He maintains that in a short time of the infant's sensory life, her retina was stimulated in the "red frequencies", her headbones, tongue, and larynx were affected in specific ways by her babbling something that sounds similar to "Red!", and, most importantly, there were some impacts on the infant's nerve endings that made the episode pleasant. In [3], the consequences of this overall sensory experience for the infant's acquisition of her first language are delineated. On another occasion where a red thing is present in the relevant sense, her innate standards of perceptual similarity ensure that she will perceive this situation as similar to the one involving the babbling of "Red!". As this latter episode was pleasant, the child strives to further increase the former's similarity to the latter one, which leads her to utter "Red!" again. If all goes well, the parent will once again reward this utterance of the infant, since there is again a red thing present in the relevant sense.

The notion that I want to focus on is the one of *pleasure through approval.* In the paradigmatic behavioristic setting, the reward would consist in a pellet of food distributed by a mechanism to a small animal, say a rat, compare Quine (1974, pp. 16, 29). However, in the passage above, Quine explicitly includes the option of pleasure through approval.

The pleasure that the infant gains from her parent's approval of her utterance is substantially different from the pleasure of a rat getting its pellet of food. I focus on two important disanalogies. The first one is that approval is not a physiological or a physical kind. The pleasure gained through approval is not determined and thus not predictable by any physical properties of the means by which the approval is transmitted. One can easily imagine pleasure caused through approval that is conveyed by means of very unpleasant sounds (imagine a proud father sounding a terribly loud and high-pitched horn at the arrival of his daughter at the finish line of her first marathon).

I now turn to the second and more important disanalogy between being pleased by a pellet of food as opposed to being pleased by approval. Consider the following statement, with which the father approves of his daughter's first utterance of "Red!" in an appropriate situation:

4. *This is correct, my daughter.*

Let us see how this expression of approval would be cashed out in Quine's behaviorism. As is the case for any sentence, the father must have acquired a disposition to respond to a certain class of stimuli with this utterance. Now, a first problem already emerges: the daughter is supposed to have uttered "Red!" for the first time in correct circumstances, hence the father's utterance of (4) in response to this first correct utterance cannot be guided by such a disposition. But let us grant that this problem can somehow be accounted for. Hence, from now on, whenever the stimulus is given, that is, whenever the daughter utters "Red!" in the correct circumstances, the father utters (4).

Now, let us move on to the infant's side of this situation. For her to acquire a disposition to utter "Red!" whenever something red is present, she must be rewarded for her correct utterances to acquire the respective disposition. Thus, (5) must be true.

5. *The child is pleased by the father's approval of her utterance.*

(5) necessarily transcends the behavioristic setting in which Quine wants to set up his story of first-language acquisition. It takes quite something to be pleased by approval. The infant has to conceive of her utterance, say "Red!", as something that is subject to normative evaluation: she must see

it as something that can be approved of or disapproved of according to some standard or other. Also, the infant has to be aware that the father's utterance is such an expression of approval, that is, that it is *about* her own utterance, that it has the *purpose* to convey a normative evaluation of her utterance. This awareness requires that the infant conceives of her parent as a subject capable of making such evaluations, and of the utterance itself as conveying such an evaluation.

Finally, it must find pleasure *in the mere fact* that the parent is approving of her utterance. This kind of pleasure is not directly connected with the bodily needs of the infant, but rather with her needs as a communicative being. In short, crediting the infant with the necessary abilities to be able to be pleased by approval involves all of the notions that Quine puts on the bad, intentional side of his divide, as he sketches it in the passage quoted above (Sect. 2.1.2, p. 33): the infant must conceive of her father as minded, of her father's utterances as having a purpose, and of that purpose as not being a sober scientific statement but rather a normative judgment of her own utterance.[9] I call these abilities that the infant must possess in order to be pleased through approval 'intentional abilities'.

Quine's use of pleasure through approval in *The Roots of Reference* is not a one-time slip of tongue. Already in *Word and Object*, he writes that repeating 'Mama' when somebody utters the same word "will continue to be rewarded, for everyone will applaud the child's seeming mimicry" (Quine 1960, p. 81). Again, Quine conceives of *applause* as a type of reward. In strict analogy to pleasure through approval, this notion of reward through applause presupposes intentional abilities on the part of the individual.[10]

[9] For completeness' sake, note that 'to approve' can be used opaquely, as in 'Peter approves of a right to asylum for all citizens of countries that are at war.' This statement could be true even if, happily, no country would be at war at the time Peter utters the statement.

[10] A similar problem arises in Quine's discussion of radical translation. As Glock (2003, pp. 175–182) shows, Quine's conception of the situation in which radical translation occurs is so radically behavioristic that the radical translator would have virtually no hope of getting his translation studies started without some "closet hermeneutics" (Glock 2003, p. 175). In particular, Glock argues that, *pace* Quine's official doctrine, his account presupposes that the native and the translator "engage in a specific kind of dialogue, that is, perform certain types of speech acts" (Glock 2003, p. 179). Among others, Glock points out, Quine presupposes that the native understands that the conversation with the translator has a specific aim, namely teaching and learning a language, respectively.

This intentional conception of pleasure, however, seems to pose a problem for the very project in which Quine is engaged. Put slightly polemically, one only buys into his behavioristic account of how the infant acquires her first observation sentences by failing to notice an ambiguity in the notion of pleasure at work. This ambiguity is ultimately a consequence of transplanting the conceptual apparatus of behaviorism from the rat laboratory to the infant at home acquiring her first language.

A conceivable objection to this line of reasoning is that this critique of Quine takes some of his assertions too literally, while it would be possible to reinterpret these assertions so as to make them consistent with his mechanistic-behavioristic principles. As Burge (2010, p. 116) may put it, the critique unnecessarily hyper-intellectualizes the notion of approval in play.

Could it not be that the parent's expression of approval has a non-intentional, purely physiological side effect that is consistent with Quine's behaviorism? For instance, in principle, the reward could consist in the purely acoustic properties of expressions of approval, as opposed to expressions of disapproval. Infants could simply like the sound of 'yes' and dislike the sound of 'no'. This, however, would have rather absurd consequences. First, we would have to postulate some perhaps innate and evolutionarily selected mechanism that ensures that the infant finds precisely those rather drastic acoustic variations of 'yes' pleasing as they occur in English, and Spanish, and Arabic, etc. Second, the infant could reward itself as much as she wanted by simply uttering 'yes' very often and forgetting about the world around herself. This evinces that the reward cannot lie in the sound waves caused by the parent's 'yes', as they reach the infant's eardrums.

But maybe some intersubjective, but still clearly non-intentional phenomenon could serve as the reward? Perhaps just the parent's making a happy face, or focusing on the infant? The idea would then be that the infant acquires the disposition to utter "Red!" in just the right situations because in these, and only in these, the parent reacts to it in a way that is intrinsically pleasing to the infant, while still being clearly non-intentional. For instance, little Paula goes for a walk with her parent and comes across a red umbrella. Paula utters "Red!", and the parent expresses her approval with great enthusiasm. Now, Paula doesn't care about the approval—she doesn't even know what that would be. But she has an innate craving for the parent's attention, as expressed, say, through eye contact, and that craving is now satisfied.

This is probably the most promising way for the Quinean to deal with this issue, but it probably fails. On this conception of reward, the infant would lose her disposition to utter the observation sentence in the correct circumstances soon, since she would soon stop to be rewarded by the parent in the relevant way. At the time where the infant has just learned "Red!", she may receive an enthusiastic sort of attention, or smiling—but for how long? One week, two weeks? After some time, depending on the patience of the parent, the trick would stop to work for the infant: the parent would react to the infant's utterance either with merely intentional approval (something like "Yes, that's also red"), while going about her business. After some time, the parent may even react with impatience, since it gets annoying to be reminded of every single bit of red that is around. The infant's parent is no mechanical apparatus that rewards the infant always and only if the infant utters the right sentence in the right situations.

More generally speaking, without pleasure through approval, the infant is in a hopeless situation when learning her first observation sentences: there are not enough *nonintentional* rewards for the infant to develop the drive necessary to acquire *and sustain* a disposition, say, to respond with "Red!" to every presentation of a red shawl. The problem of the scarcity of non-intentional rewards is seriously aggravated by the gigantic dimension of the infant's task, as Quine conceives it: remember that the infant must acquire a new disposition for each new sentence-situation-pair that she learns—and has to be continuously rewarded to acquire and sustain this myriad of dispositions. For instance, she needs one disposition to correctly respond (with "Red!") to a red ball, another one to correctly respond to a red shawl, yet another one to correctly respond to a red curtain, etc. As it were, this is the poverty of the stimulus problem on steroids.

Therefore, once the rewards are thoroughly de-intentionalized, it is highly implausible that the infant will ever learn the correct use of just a small number of observation sentences in just a small number of different situations, no matter how harmonized her similarity standards are with the parent's. At best, the infant would come up with a correct use of an observation sentence in a given situation, be non-communicatively rewarded for it for some time, then stop using it correctly for lack of non-communicative reward. Then, come up with the correct use of another observation sentence, etc.

If this is right, then Quine is confronted with a difficult choice. He can either take back the idea that the pre-linguistic infant can be pleased by the parent's expression of approval, or applause. But then his entire account of how she learns her first observation sentence becomes highly implausible. Or Quine can affirm that the infant is indeed capable of being pleased through the parent's approval of her expressions; but this decisively contradicts both Quine's behaviorism and his physicalism.

Let me sum up these two internal criticisms of Quine's mature naturalized epistemology. The first of them concerns Quine's invocation of natural selection to explain why communication is possible and why our science is successful, and it is based on methodological considerations of evolutionary biology: the information required for such an evolutionary explanation has not been provided by Quine, and it probably cannot be provided.

In this section, I have shown that Quine's accounts of the infant's first steps into language attribute the ability to understand communicative purposes, that is, a kind of final cause, to the infant. Furthermore, I have suggested that, once Quine's account is purged of this ability, it loses any plausibility, since the infant then faces a task that it probably cannot accomplish. This present issue threatens to undermine Quine's ability to explain why we have any language, and hence any science at all.

To conclude, note again that Quine's eschatological physicalism is seriously questioned by the worries detailed in the present section. Indeed, the younger Quine may have decided to abandon this doctrine entirely. This is at least suggested by the following passages from the 1940s (*On What There Is* was first published 1948), where Quine proposes an experimental attitude to see how far we can take both the physicalistic and the phenomenological conceptual scheme:

> Here we have two competing conceptual schemes, a phenomenalistic one and a physicalistic one. Which should prevail? Each has its advantages; each has its special simplicity in its own way. Each, I suggest, deserves to be developed. Each may be said, indeed, to be the more fundamental,

> though in different senses: the one is epistemologically, the other physically fundamental. (Quine 1980 [1948], p. 17)
> But the question what ontology actually to adopt still stands open, and the obvious counsel is tolerance and experimental spirit. (Quine 1980 [1948], p. 19)

Unfortunately, that experimental spirit, indeed, the iconoclast from *Two Dogmas*, seems to have gotten lost somewhere along the way.

References

Brandon, Robert N. 1990. *Adaptation and Environment.* Princeton: Princeton University Press.

Burge, Tyler. 2010. *Origins of Objectivity.* Oxford: Oxford University Press.

Føllesdal, Dagfinn. 2014. Developments in Quine's Behaviorism. In *A Companion to W.V.O. Quine*, ed. Gilbert Harman and Ernie Lepore, 263–278. Chichester: Wiley Blackwell.

Glock, Hans-Johann. 2003. *Quine and Davidson on Language, Thought and Reality.* Cambridge: Cambridge University Press.

Kornblith, Hilary. 1993. *Inductive Inference and Its Natural Ground.* Cambridge: The MIT Press.

———. 2002. *Knowledge and Its Place in Nature.* Oxford: Clarendon Press.

———. 2014. Knowledge in Humans and Other Animals. In *A Naturalistic Epistemology: Selected Papers*, 119–141. Oxford: Oxford University Press.

Maddy, Penelope. 2017. *What Do Philosophers Do? Skepticism and the Practice of Philosophy.* The Romanell Lectures. Oxford: Oxford University Press.

Plantinga, Alvin. 2002. The Evolutionary Argument Against Naturalism. In *Naturalism Defeated? Essays on Plantinga's Evolutionary Argument Against Naturalism*, ed. James Beilby, 1–12. Ithaca: Cornell University Press.

Quine, Willard Van Orman. 1960. *Word and Object.* Cambridge: The MIT Press.

———. 1969. Natural Kinds. In *Ontological Relativity and Other Essays*, 114–138. New York: Columbia University Press.

———. 1974. *The Roots of Reference.* LaSalle: Open Court Publishing Co.

———. (1980 [1948]). On What There Is. In *From a Logical Point of View*, 1–19. Cambridge: Harvard University Press.

———. 1992. *Pursuit of Truth*, 2nd ed. Cambridge: Harvard University Press.

———. 1995. *From Stimulus to Science*. Cambridge: Harvard University Press.
———. 2016 [1995]. Response to Gary Ebbs. In *Quine and His Place in History*, ed. Frederique Janssen-Lauret and Gary Kemp, 33–36. Basingstoke: Palgrave Macmillan.
Richardson, Robert C. 2007. *Evolutionary Psychology as Maladapted Psychology*. Cambridge: The MIT Press.

4

Burge: Proto-Predicates in Perceptual Representations

In this chapter, I discuss Tyler Burge's account of the origins of objectivity. According to him, perceptual representation has a proto-predicational structure, is achieved even by non-mammals and studied by a bona-fide scientific discipline. As mentioned in the introduction, I discuss Burge's proposal, as it were, with an eye looking backwards and an eye looking forward. First, Burge's account suggests that Quine's mature proposal has been superseded by the progress of science. Second, I consider the merits of Burge's account as an attempt at naturalized epistemology in its own rights. In the remainder of this introduction, I elaborate on this double significance of Burge for naturalized epistemology.

First, note that Burge does not see his work as a contribution to naturalized epistemology. Indeed, he explicitly rejects projects like Quine's that try to pursue psychology without making use of irreducibly mental notions like *representation* (Burge's concept of naturalism, as he uses it in the following passage, is strongly tied to ontological forms of reductionism):

R. Gubelmann, *A Science-Based Critique of Epistemological Naturalism in Quine's Tradition*, https://doi.org/10.1007/978-3-030-24524-5_4

> I find such rescue missions in philosophy [attempts to reduce psychological notions to non-psychological ones, RG] quaint and the parallel moves in psychology retrograde. Each approach is out of sync with empirical knowledge and practice. Neither approach captures, or even reflects on, what is distinctive about certain powerful and successful explanations in psychology, specifically perceptual psychology. (Burge 2010, p. 296)

In addition to stating his rejection of projects like Quine's, Burge here also gives a reason for this rejection that will be important towards the end of this section: he argues that the basic problem with such 'rescue missions' is that they ignore certain powerful explanations in psychology. In other words, the basic objection to projects like Quine's eschatological physicalism is that they disregard successful scientific theorizing—for a naturalist, this should be a serious objection.

This provides the first goal that I am pursuing with my discussion of Burge: If sound, Burge's claims provide a strong, science-based critique of Quine's naturalized epistemology. Contemporary perceptual psychology is of obvious importance for Quine's naturalized epistemology, given Quine's reliance on innate perceptual similarity standards in his mature naturalized epistemology. Very briefly (see below, Sect. 4.2), Burge claims that this discipline irreducibly posits intentional representations, that it employs a kind of teleology that is strictly distinct from the only teleology that Quine is prepared to accept, namely conduciveness to survival, and that the perceptual representations posited by perceptual psychology are structured in a singular and a general part. These three claims directly contradict basic tenets of Quine's naturalized epistemology.

The second and more constructive goal is to assess whether Burge's reading of perceptual psychology can be seen as a successful attempt at naturalized epistemology in itself: could it be that perceptual psychology has succeeded in scientifically understanding and explaining these cognitive structures—proto-propositional representations in perception—that are essentially similar to the fully-fledged propositions, as they are used in scientific cognition? In this case, Burge could be seen to advance Quine's project of naturalized epistemology, even if he would do so in a way that clearly differs from Quine's original proposal.

Following Burge (2010, pp. 36–37), I use the 'as of'-terminology: a representation of a duck is a representation whose distal stimulus is a duck. A representation as of a duck is a representation that can have anything as its distal stimulus (in principle), but whose content involves the property *duck*. Hence, a representation of, and as of a duck is a representation that accurately represents a duck as a duck.

In the following, I introduce contemporary perceptual psychology in general and its explanation of perceptual constancies in particular (Sect. 4.1). Then, I reconstruct Burge's threefold conception of objective representation, and I argue that Burge is right to claim that this conception is implicit in successful theories of perceptual psychology (Sect. 4.2). Finally, I discuss whether the Quinean can ignore contemporary perceptual psychology, much like, as I urged in the preceding chapter, she can ignore nativism that is based on the habituation and preferential looking paradigm (Sect. 4.3).

4.1 Contemporary Perceptual Psychology

In this section, I briefly delineate the discipline of perceptual psychology in general. Then, I focus on its explanations of perceptual constancies.

Contemporary perceptual psychology, as pioneered by David Marr (2010 [1982]), stands in an interesting relationship to Quine's doctrine of the inscrutability of reference. Perceptual psychology conceives of perception as consisting of a number of informational problem-solving processes, the most general problem being underdetermination of the distal stimulus through the proximal stimulus.

From a Quinean perspective, this general problem-setting is interesting. It's accurate to say that Quine's *modus ponens* to infer the inscrutability of reference is perceptual psychology's *modus tollens*. Below is how one can conceive of Quine's epistemic argument for the inscrutability of reference[1]:

[1] I am not claiming that this epistemic argument is the only argument given by Quine in support of this. I merely claim that this is one of the arguments that Quine indeed proposed. For two passages where Quine deduces the inscrutability of reference epistemically, consider Quine (1981, pp. 21–22) and Quine (1992, pp. 31–34). Compare below, Sect. 7.3.2 for an extended discussion of Quine's doctrine of the inscrutability of reference.

P1 The proximal stimulus available to humans (and other species) underdetermines central features of the corresponding distal stimulus.

P2 Humans (and other species) do not have any other source of information than the proximal stimulus, based on which they could determine central features of the corresponding distal stimulus.

K Hence, humans (and other species) are unable to determine central features of the distal stimulus corresponding to a given proximal stimulus.

Quine's argument is subtler than this, but this three-liner captures the gist of his epistemic argument: patterns of sensory stimulations of our nerve endings do in many ways underdetermine the distal stimulus, typically a middle-sized object. Furthermore, we do not have anything else to go on with than these patterns of sensory stimulation (remember Quine's source-empiricism, as detailed above, in Sect. 2.1.4). This implies that humans are unable to determine central features of the distal stimulus unequivocally; that is, there are always alternative conceptions of the distal stimulus that also fit any available evidence. In this sense, the distal stimuli to which our sortal terms refer are inscrutable.

Perceptual psychology, as it is currently practiced by most researchers in the field, accepts Quine's *P1*: the proximal stimulus does in fact not determine central features of the distal stimulus. However, it rejects the conclusion: perceptual psychologists presuppose that what they call the visual system typically manages to overcome the underdetermination of the distal stimulus by the proximal stimulus. Thus, the visual system is able to reliably produce an accurate representation as of the distal stimulus. Hence, perceptual psychologists conclude, Quine's *P2* must be wrong. The structure of this argument is as follows:

¬K Humans (and other species) are typically able to determine central features of the distal stimulus corresponding to a given proximal stimulus.

P1 The proximal stimulus available to humans (and other species) underdetermines central features of the corresponding distal stimulus.

¬P2 Humans must have more than the proximal stimulus to determine central features of the corresponding distal stimulus.

Here, *P2* is highly unspecific: what exactly does this 'more than the proximal stimulus' amount to? Contemporary perceptual psychologists, led by Marr (2010 [1982], pp. 22–27), have decided to cash out this 'more than the proximal stimulus' by positing algorithms that aim at, and typically succeed in, producing representations that accurately represent the distal stimulus. Here, 'algorithm' refers to a set of rules that are implemented one way or another in the human visual system. These rules specify on a generic level how the visual system approaches specific information processing tasks. Such algorithms could in principle as well be implemented in a suitable computer. These algorithmic rules are often symbolic, that is, they take intentional entities, representations, as input and deliver other intentional entities, representations, as output.

Furthermore, it is worth emphasizing that this basic conception of perception cannot dispense with representations (I will defend this claim in more detail below, Sect. 4.2). If it is true that the distal stimulus is underdetermined by the proximal stimulus, then the only way for the algorithms to overcome underdetermination is by producing something distinct from the proximal stimulus that represents the distal stimulus.[2]

In summary, contemporary perceptual psychology is genuinely psychological insofar as it conceives of perception as an information-processing

[2]When it comes to early vision, the so-called direct approach to perception, pioneered by J. J. Gibson, avoids both the fundamental problem of underdetermination and the need to posit any representations to overcome this underdetermination. For a recent defense of this approach, compare Warren (2012). According to direct perceptionists, the central problem of contemporary perceptual psychology, namely how to overcome the underdetermination of the distal stimulus by the proximal stimulus, is a consequence of a misguided conception of perception. Properly conceived, there is enough information in the so-called visual array, in the way a situation presents itself to an animal that looks at it from different perspectives, for some time, to simply see things, without having any need for formation principles to overcome any underdetermination.

Color constancies could then, according to the Gibsonian, still be the product of intricate physiological systems, to be investigated and elucidated with great profit by physiology, but these systems would simply be psychologically irrelevant. What matters psychologically is that the animal is seeing the same color under different light conditions. Perceptual psychologists in the tradition of Marr would reply that they are explaining, and explaining psychologically, what the Gibsonian is simply accepting as a brute given, to wit, the animals' ability to discriminate luminance edges from reflectance edges. Direct perceptionists confirm my basic claim insofar as they are forced to deny underdetermination to be able to avoid positing representations.

task, centrally involving intentional entities, namely representations, that are produced and manipulated by algorithmic processes. Furthermore, generally speaking, the central information-processing task of the visual system is to overcome the underdetermination of the distal stimulus by the proximal stimulus. This conception of perception as having tasks to accomplish brings with it an inherent teleology: perceptual systems are designed to optimally solve these tasks.

In the remainder of this section, I focus on the examples of lightness and color constancies to show how well-understood these constancies are by perceptual psychology and how detailed the empirical support is that they can gather. Throughout this section, I am relying on one of the most established textbooks in the field, Palmer's *Vision Science*. To illustrate and corroborate the account given by Palmer, I occasionally refer to relevant research articles.

Lightness Constancy in Us. Let me sketch the basic problem of lightness constancy.[3] Take your book from the study, where the main source of light is a desk lamp, out into bright sunlight. In the study, the difference in terms of light intensity impacting on the retina from black letters as opposed to from the white background is about 1:9: the white pages reflect about nine times more light than the black letters. This difference is enough for us to see the letters as clearly black and the pages as clearly white. In contrast, out in bright sunlight, the light intensity, issuing forth from the black printed letters on the book's pages and impacting on our retina is about 100 times higher than the one issuing forth from the white pages in the study. Hence, if it was only the proximal stimulus that would determine lightness perception, we should all see the black letters as white in sunlight. However, thanks to lightness constancy, we all perceive the letters as black in both conditions, in spite of these drastic differences in proximal stimulation. This is lightness constancy at work (compare Palmer 1999, p. 125).

More technically speaking, the task to be accomplished by the visual system in the case of lightness constancy is as follows. The proximal stimu-

[3] A terminological remark: lightness constancy is often also called 'achromatic color constancy', and sometimes called 'brightness constancy'.

lation only affords information regarding luminance, that is, regarding the intensity with which the photoreceptors in the eye are stimulated. This underdetermines the distal cause of this proximal stimulus because the intensity of this luminance is the combined result of the surface reflectance (roughly, the color of the surface) and the illumination of the relevant surface (roughly, the light intensity impacting on the surface, Palmer 1999, pp. 126–127). This means that there are always two factors involved in causing the luminance, that is, the intensity of stimulation of the photoreceptors. These two factors are surface reflectance and illumination of the surface. The task of the visual system is to reliably discriminate between the part of the luminance that is merely due to the illumination of the surface and the part of the luminance that is due to an actual property of the distal stimulus, namely its surface reflectance. Since it is obvious that the human visual system *can* accomplish this, the remaining question is *how* it achieves this.

According to Palmer (1999, pp. 128–133), the visual system employs a number of rules (in the sense of algorithmic rule) to distinguish between illumination and reflectance; I here focus on the two that are most important. First, Palmer (1999, p. 127) claims that "perceived lightness depends on the *relative luminance* (or **contrast**) between neighboring regions". That is, put simply, something will look white to us if it is adjacent to areas that cause a much lower luminance on the photoreceptors, and, conversely, something will look black if it is adjacent to areas that cause a much higher luminance on the photoreceptors. While the absolute amount of luminance caused by the black letters illuminated by sunlight differs drastically from the absolute amount of luminance caused by the black letters illuminated by the desk lamp, the relative luminance between the black letters and the white background pages stays the same. Hence, assuming that our visual system achieves lightness constancy at least partly with recourse to ratios of luminance can contribute to explaining lightness constancy.

But how precisely does the visual system compute these luminance ratios? With regard to this question, Palmer (1999, pp. 128–129) relates research that suggests that only the luminance ratios at the edges matter. This means that when our visual system computes the reflectance of two surfaces, their relative reflectance is computed solely based on the reflectance changes right at the respective edges. Hence, the rule followed by the visual system could be put as in rule 1.

Rule 1. *If the luminance edge on the retina is sharp, represent a reflectance edge; otherwise represent an illumination edge.*

To illustrate this rule 1, it is best to use an example: The Craik-O'Brian-Cornsweet illusion. In this illusion (compare Palmer 1999, p. 128), we are presented with a display that contains two rather dark areas with a significantly brighter area between them. The transition from one of the darker areas to the lighter area is significantly sharper than the other. Our visual system represents the area with the sharp transition as slightly darker than the one with the more graded transition. However, as a matter of fact, the two sides have the same reflectance, apart from a small zone around the luminance edge.

Experiments such as this one have been used by perceptual psychologists to determine the general algorithm that our visual system uses to represent constant reflectance (roughly, color) of the distal stimulus in spite of stark differences in the proximal stimulus.[4]

What rule 1 does not determine, however, is how our visual system maps the *relative* information obtained by following this rule on the *absolute* scale of black and white. Surprisingly, there is experimental evidence that, for any given scene, the visual system simply represents the surface that causes the highest luminance intensity on the retina as white. In an experiment conducted by Gilchrist et al. (1999, pp. 799–800), the subjects were shown nothing else than a large acrylic hemisphere which was divided in semispheres, one of which was black, the other gray. With overwhelming agreement, the subjects did report the dark gray as white and the black as middle gray, thus confirming the simple heuristic. Hence, the second rule can be specified as in rule 2.[5]

---

[4]Note, however, that more specific aspects of how our visual system achieves lightness constancies have been explored in substantial mathematical detail since the late sixties, compare, for example, Robert B. Freeman (1967).

[5]Note that the issue here is only achromatic color; hence, rule 2 does not imply that every scene that we perceive has to have white in it.

Rule 2. *Represent the brightest area in a given perception as white.*

Color Constancy in Honeybees. Color constancy is a more complex ability than lightness constancy. As Palmer (1999, p. 133) details, color constancy is needed because light from different sources, say, from the sun versus from a light bulb, has different illumination spectra and hence causes different luminance spectra on the eye. This means that, without color constancy, we would see a different color when taking, say, a blue ball from the study out into the bright light of day.

Regarding color constancy, I would like to focus less on the algorithm that perceptual psychology postulates to explain this ability, and more on the question which non-human species possess it, and how vision scientists establish this fact. Chittka et al. (2014) provide an extensive overview on the field of bee color constancy. They (ibid., pp. 438–439) report on a first experiment in the sixties by Georgii A. Mazokhin-Porshnyakov that runs as follows. Bees were trained to yellow. This means, the bees were repeatedly presented with a plane made up of squares of different shades of gray, except for one square that was yellow. The experimenter put a reward on the yellow square, namely sugar solution (hence, the set-up looks somewhat like in Fig. 4.1). After repeatedly presenting the bees with this set-up, the bees instantly flew to the yellow square whenever presented with the display and got their sugar, whereas before training, they had examined the different areas without a clear pattern.

Mazokhin-Porshnyakov then tested whether the bees still immediately choose the yellow square if they are also given the possibility to choose a gray square that gets its daylight filtered through a yellow filter such that its physical spectral properties match exactly with the original yellow one. In other words, in this set-up, the proximal stimulus that the bees experience from the yellow square is physically identical to the proximal stimulus that the bees experience from the light-filtered gray square.

This indistinguishability of the two proximal stimuli notwithstanding, in the experiment, the bees clearly preferred the original one. That is, they clearly preferred the square with the yellow reflectance, as opposed to the square with the yellow illumination. This means that the bees'

Fig. 4.1 The set-up for Mazokhin-Porshnyakov's experiment, with the deviation that the squares that are black or white here were different shades of gray in his experiment (*Source* Author)

visual systems were able to distinguish the surface property of the square, its reflectance, from the illumination of the surface. If the visual systems weren't able to make this distinction, approximately fifty percent of the bees would have landed on the yellow square and fifty percent of the bees would have landed on the yellowish-illuminated gray square (as the proximal stimulus of both squares was identical).

A further study referred to by Chittka et al. (2014, p. 439), conducted by Werner (1990) considerably refined this basic paradigm. The set-up was a screen divided up into squares, or plates, in the manner of a chessboard, but with the white squares replaced by colored filters. Behind the screen were three different lamps that could be separately dimmed to create a broad range of light spectra and intensities. The experiments were conducted in a room lit solely through the lamps of the set-up, hence no interference of other light was possible. The bees were then trained to land on a patch of specific reflectance, say, yellow. Then, the illumination was changed such that a different plate, called the matching plate, with different reflectance, caused exactly the same proximal stimulations of the honeybees' eyes as the proximal stimulations that had been caused by the training plate in the original illumination. Under the new illumination, the training plate itself did cause different proximal stimuli than under the original illumination. In this experiment, too, bees landed on the training plate much more often than on the matching plate. This behavior demonstrates that their visual systems are able to distinguish between the reflectance and the illumination of the respective plates.

Note that, in this last experiment, the behavior of the bees is exactly the opposite of what one would expect if they navigated based on proximal stimuli only. Then they would have chosen the matching plate, since this plate caused exactly the same proximal stimuli as the training plate.

In sum, in this section, I have shown that contemporary perceptual psychology accepts Quine's claim that the proximal stimulus underdetermines the distal stimulus. However, contrary to Quine, perceptual psychology assumes that our visual systems are typically able to overcome this underdetermination. Our visual systems achieve this with recourse to rule-governed processes that produce and manipulate representations as of the distal stimuli. Then, I have sketched how perceptual psychology employs such rules to explain how the human visual system achieves achromatic color constancy. Finally, I have detailed the kinds of experiments that perceptual psychologists conduct to establish that the visual systems of honeybees do achieve color constancy.

4.2 Burge's Case for Objective Representation in Perception

In this section, I detail Burge's threefold conception of objective representation in perception and the justification that Burge offers for this conception.[6]

According to Burge, perceptual representations are governed by perceptual norms, they are perspective-independent, and they are structured in a particular and a general part.[7]

Objective Representation as Subject to Perceptual Norms. According to Burge, representations are necessarily subject to accuracy conditions. This is one of the central messages of Burge's *Origins of Objectivity*. These accuracy conditions are conceived as "natural norms" (Burge 2010, p. 312) that are constitutive of perceptual representation; this means that nothing can be a perceptual representational state that is not subject to accuracy conditions:

> And being inaccurate or non-veridical is a kind of failure or shortcoming for perception. Nothing could count as a perceptual state if it would not undergo a kind of failure if it were not approximately accurate. [...] Having veridicality conditions or accuracy conditions is an aspect of what it is to be a perceptual state. That is, perceptual kinds constitutively, and by nature, include veridicality conditions. (Burge 2010, p. 512)

In this passage, Burge emphasizes that perceptual states (a term that he uses interchangeably with representations in perception) are necessarily subject to accuracy conditions (Burge [2010, p. 12] uses 'accurately' and 'veridically' as well as 'accuracy conditions' and 'veridicality conditions'

[6]Note that Burge himself does not explicitly conceive of his conception of objective representation as threefold. This is a contribution of my part. Furthermore, instead of 'perceptual representation' or 'representation in perception', I usually just write 'representation', since I am almost exclusively concerned with representations in perception.

[7]Burge lists all three of these essential properties of objective representations in the following passage: "The objectivity of such representation lies not only in its accuracy and its specifying relevant attributes in a way that entails their physicality. It also lies in the physical subject matters being mind-independent, or at least constitutively non-perspectival" (Burge 2010, p. 59).

interchangeably); it is not possible for something to be a representation and not to be governed by accuracy conditions. Hence, representations presuppose a specific kind of teleology, namely perceptual teleology: representations, if successful, fulfill the function of representing accurately.

Burge emphasizes time and again that representational functions and norms are different and independent from biological (i.e. evolutionary) functions and norms. For example: "Whatever function is fulfilled in cases of representational success (and frustrated in cases of representational failure) is not a biological function" (Burge 2010, p. 308). Burge defends this distinction between perceptual accuracy and fitness advantage in two ways. The first one simply points out that biological function is not the same thing as representational function: representational success may hinder biological success, that is, confer an evolutionary disadvantage on its bearer, while representational failure may contribute to biological success. The second line of reasoning capitalizes on the role that such perceptual teleology plays in successful scientific explanation, and it will be discussed later in this section.

Objective Representation as Perspective-Independent Representation. Burge's representations are objective also in the sense that they transcend the proximal, subjective perspective of the perceiving individual and represent a distal stimulus that is, unlike proximal stimulations, equally accessible to different subjects. For example, a slanted circle, being the distal stimulus, should be represented as a circle, and not as an ellipse—even though the pattern of physical stimulation on the retina, the proximal stimulus, has the shape of an ellipse. By the same token, the perspective-independence specifies the telos of perceptual teleology: for an objective representation to represent accurately is to accurately represent *the distal stimulus.* A central notion for Burge's account of the perspective-independence of representations is *objectification*:

> Objectification, then, is the formation of a representational state that represents the physical environment, beyond the individual's local, idiosyncratic, or subjective features. (Burge 2010, p. 397)

According to this passage, *objectification* is a process that transcends the subjective, proximal circumstances of an individual and represents parts of the distal physical environment. A process that fulfills these requirements has been discussed above (Sect. 4.1): lightness constancy. Lightness constancy is the ability to represent the reflectance of a surface, that is, a stable aspect of the distal physical environment, in spite of stark changes in the corresponding proximal stimulus (Burge 2010, p. 399). For instance, the bee represents a particular distal stimulus, a surface, as having a constant reflectance in spite of drastic changes in the relevant proximal stimulus.

Objective Representation Has Proto-Propositional Structure. Burge's objective representations do not have propositional structure, but something that clearly resembles Fregean argument-function structures: objective representations always contain a particular part that refers to a given particular thing in the perceived environment and a general, repeatable part that attributes a property to the particular. Burge illustrates his notion of perceptual particulars and attributes in the following passage:

> Take a visual perceptual state as of a cylindrical solid. There are two aspects of perceptual representational content of the state – general and singular. The singular aspect functions fallibly to single out (refer to) perceived particulars. When successful, the perceptual state refers to a particular cylindrical solid, and perhaps particular instances of cylindricality and solidity. The general aspect in the representational content functions fallibly to group or categorize particulars by attributing some indicated kind, property, or relation to them. When successful, the perceptual state attributes cylindrical solidity to a particular cylindrical solid. (Burge 2010, p. 83)

According to this passage, a perceptual state, a representation, is constituted by a singular and a general aspect. The singular aspect should refer to a particular, say *this* thing there, while the general aspect has the function of grouping these referents in some class, say the class of cylindrical shapes. As Burge emphasizes in this passage, these representations are fallible, that is, they can represent accurately or inaccurately.

Furthermore, the particular part of a representation needs the general part of a representation to refer to the respective environmental particular: "One cannot perceive a particular without perceiving it by way of some general, repeatable capacity to *attribute* properties, relations, kinds, veridically" (Burge 2010, pp. 454–455). Thus, perceiving something always means perceiving something with a specific perceptual attribute, or rather, a number of them. There are no bare perceptual particulars. In a critical discussion of Pylyshyn in a footnote, Burge (2010, 455n) argues that it is not possible to perceive bare, that is, attributeless particulars because some criteria or other are needed to distinguish the particular from its environment. Burge also claims that tracking bodies does in fact require only very little of the familiar properties and attributes. But even reference to specific bodies can only be established by means of some attributes.

Finally, Burge (2010, p. 541) argues that the repeatable attributes are not conceptual because they cannot be used in *pure attribution*, that is, beyond contextual references to particulars. For Burge, this implies that perceptual representations cannot figure in propositional inferences. Hence, these representations constitute a *sui generis* level of cognitive sophistication, located between mere sensory registration and fully-fledged conceptual thought. This clearly distinguishes these particular-with-attribute structures from *bona fide* propositional structures (and it sets the structure that Burge postulates apart from the cognitive abilities that nativists attribute to infants, see Sect. 5.1).

Note, furthermore, that all of these three aspects of objective perceptual representations are diametrically opposed to Quine's mature conception of how the pre-linguistic infant is connected to her environment. First, according to Quine, the infant's perceptions, even when associated with an observation sentence, are not structured in a general and a repeatable part. Second, these perceptions do not *represent* a *distal, objective feature of the world*; rather, they *are* a salient subset of the respective *proximal stimulations*. Third, these Quinean perceptions are not governed by any natural norms.

How Perceptual Psychology Supports Burge's Conception of Objective Representation. In the following, I show that the way perceptual constancies are explained by perceptual psychology does indeed lend support to Burge's threefold conception of objective representation in perception.

According to Burge (2010, pp. 351–354), what is crucial to perceptual constancies' being vehicles for objectification is that the constancies are achieved by formation principles (Burge calls 'formation principles' what I have called algorithmic rules, see Sect. 4.1). Environmental regularities determine the character of these formation principles that in turn govern the visual system's discrimination between changes in proximal stimulations that are and others that are not indicative of changes in distal objects. In the case of color constancy, Burge spells the relevant formation principle out in the following way[8]:

> The principle centers on the degree of sharpness of the luminance contour. In the absence of contrary information, the visual system forms a perception as of a reflectance edge (rather than as of an illumination edge) in response to a *sharp* luminance contour. (Burge 2010, p. 353)

According to the principle that Burge specifies in this passage, the visual system achieves color constancy by means of the criterion of *sharpness*: if the luminance contour is sharp, the visual system produces a representation as of a reflectance edge, that is, again roughly speaking, the perception as of a change in color. If the luminance contour is not sharp, the visual system represents a change in illumination, that is, a change of the light that falls on the surface in question. Burge (2010, p. 354) emphasizes that this criterion of sharpness is only part of the overall algorithm that the human visual system uses to distinguish reflectance from illumination of a surface. This Burgean formation principle agrees nicely with what I have called rule 1 above.

Positing a representation is unavoidable in this explanation, given that the formation principle is said to form something that is strictly distinct from the sensory registration, that is, the proximal stimulus, on the retina.

---

[8] Burge's introduction of perceptual psychology's explanation of lightness constancy and the relevant concepts agrees with my account (compare above, Sect. 4.1 and Burge 2010, pp. 351–354). A *luminance contour* is a change in the registration of light intensity by adjacent photoreceptors in the eye. That is, a luminance contour is a pattern of proximal stimulation of nerve endings in the eye. This proximal effect can have two systematically different distal causes, namely a sharp change in the *reflectance* or in the *illumination* of the respective surface. Roughly speaking, the former are caused by the surface's having different colors, while the latter are caused by the surface's being exposed to different spectra and intensities of light, for example, because part of the surface is in the shadow. Thus, to achieve lightness constancy, the visual system's main challenge is to distinguish between luminance contours that are caused by luminance edges and others that are caused by reflectance edges.

The luminance contour, that is, the proximal stimulus, does not only underdetermine the representation that is formed, it can even be the case that much brighter light impacting on the retina can give rise to a representation of the same or even a darker color. Furthermore, the 'something' that the visual system forms represents the distal property, namely reflectance, which means that hardly anything else than a representation could fit the bill here.

Burge emphasizes these formation principles of perceptual constancies because he thinks that they systematically distinguish between the subjective, perspective-dependent proximal stimulus and the objective, non-perspectival distal stimulus. As Burge (2010, p. 399) puts it:

> Perceptual constancies are capacities for *objectification.* Objectification separates registration of surface stimulation that is local to individual and occasion from elements in that registration that are (according to formation patterns) representationally specific to attributes in the physical environment. Thus objectification separates local, idiosyncratic registrations from representations of individual-independent, occasion-independent, mind-independent, perspective-independent reality, beyond the individual. (Burge 2010, p. 399)

Furthermore, Burge (2010, pp. 396–398) rightly urges that the representations produced by processes that are governed by these formation patterns, or principles, are necessarily structured into a particular and a general part. The basic idea is the following (my example). Consider a bee whose visual system is able to hold the color of a given flower constant in situations of vastly different illumination. It seems obvious that the representation produced by the visual system according to the formation principles needs to contain a general, repeatable part, what Burge calls 'attribute': unless the bee's visual system does represent the relevant aspect of the distal environment, namely the surface reflectance or color, it cannot be said to achieve color constancy. Furthermore, this aspect has to be repeatable insofar as the bee must be able to attribute it to different particulars, both during training and, more importantly, during the test cases. Similarly, the representations of the bee must contain, in addition to a general, repeatable part, also a particular part, referring to a specific place in the bee's perceived environment. Otherwise, the bee would not know where to fly to get the sugar—regardless of how constant her perception of the surface reflectance of either the training or the test plate is.

Finally, Burge (2010, pp. 395–396) rightly emphasizes that these explanations of perceptual constancies by perceptual psychology make non-trivial, essential use of accuracy conditions: the representation does precisely not simply co-vary with the proximal stimulus. Rather, the essential relationship exists between the representation and the *distal* stimulus, and it is a relationship of accuracy or inaccuracy.

To conclude this section, I now consider the worry that there is something about contemporary perceptual psychology as a whole that undermines the epistemic standing of its central posit, representation. Of course, Quine would raise such an objection: by positing representations, intentional entities that are irreducible in principle to physical mechanisms, perceptual psychology violates the ontological authority of physics; it is therefore unclear whether perceptual psychology would even be counted as a science by Quine.

To understand Burge's response to this worry, take a perceptual representation as of a certain predator type that is ascribed by perceptual psychology to individuals of species *S*. Let us assume that evolutionary biology has asserted that representing this predator type accurately confers a survival advantage on its bearer (to assert this is not simple at all, as I have argued, based on Richardson 2007, see Sect. 3.3.1). In this scenario, metaphysicians that reject any non-evolutionary teleology could embrace the teleology involved in claiming that the representation can be accurate or inaccurate, since *accurate* can, on a basic level, be cashed out as *fitness-enhancing*, and *inaccurate* can be cashed out as *fitness-decreasing*. Evolutionary teleology can thus figure as a sort of shadow-teleology that grounds talk of accurate or inaccurate representation that would be illegitimate otherwise.

However, positing such a shadow-teleology is in general not possible for representations postulated by perceptual psychology. First, consider an animal species *A*, individuals of which are in general very bad at accurately representing individuals of a species *B* that preys on them. Furthermore, a subset of all individuals of *A* has a mutation such that their perceptual representations as of *B*s are almost always accurate. This has the consequence that the mutant *A*s waste less energy running away from purported predators *B* that are in fact not predators. Unfortunately, it also has the consequence that the mutants are physically less fit than the non-mutants—because they only run when there is a real *B* present—which means that they have a slightly higher chance to be actually caught and killed by a real *B*. Fortunately, the selective pressure on the mutants is not strong enough to go extinct, so they persist in the overall population of *A*s. In this example, the function that perceptual psychologists (as Burge conceives them) would attribute to the mutant-*A*s' representation as of *B* would be exactly the opposite of the function that evolutionary biologists would attribute to the same representation: according to perceptual psychology, the representation's function would be to represent accurately, and not inaccurately, while evolutionary biologists would say that its function is actually to represent inaccurately, since this confers a fitness advantage on its bearers.

Secondly, the most common case for a representation that is postulated by perceptual psychologists would simply be the one where evolutionary biologists lack any necessary data (see above, Sect. 3.3.1) to determine whether representing the respective type accurately conferred a fitness advantage on the relevant individuals at the relevant time and place. Strictly speaking, the biological reductionists would then have to forbid psychologists to postulate representations in these cases.

This is precisely what Burge (2010, pp. 296–298) thinks gets things backwards: if perceptual psychology posits such-and-such and is explanatory successful in doing so, this outweighs any metaphysical concerns about the integration of the explanation's posits in some pre-conceived basic structure of the world. Hence, any worries that representational function fits ill with our contemporary view of nature are outweighed by the success of explanations that make essential use of such functions.

I do not see how a naturalized epistemologist like Quine could resist Burge's reasoning at this point. If it is the case that a scientific discipline delivers powerful, empirically well-supported explanations by positing a certain class of entities that fit ill with very general, partly speculative conceptions of the kinds of entities that are allowed in science, the powerful explanations should always outweigh the general, partly speculative worries.

Let me sum up the content of this section. I have first shown that Burge's conception of perceptual representation is threefold. First, perceptual representations are governed by accuracy conditions. Second, perceptual representations represent a subject matter independently of many of the subjective features of the representing individual. Third, perceptual representations represent particulars as having certain repeatable attributes.

Then, I have argued that perceptual psychological research on lightness and color constancies, as I have reviewed it above (Sect. 4.1), employs this threefold conception: the relevant explanations of perceptual psychology are committed to crediting the respective visual systems with representations in Burge's threefold understanding.

Finally, I have discussed an objection that concerns the epistemic standing of perceptual psychology as a whole. According to this objection, perceptual psychology is problematic because it posits intentional entities, representations, that are governed by a strictly nonbiological teleology. I have shown that Burge replies to such worries by emphasizing the successful explanations provided by perceptual psychology, and by urging that this success outweighs any general, metaphysical worries about the explanations' central posit.

4.3 Why Even Quine Should Take Perceptual Psychology Seriously

In the preceding section, I have argued that perceptual psychology does indeed presuppose Burge's threefold conception of objective representation, and I have elaborated on how Burge counters general worries regarding the epistemic standing of this discipline. This section focuses on ways of the Quinean to argue that contemporary perceptual psychology can be ignored for the purposes of his naturalized epistemology. I will argue that there are good reasons why the Quinean cannot ignore the research from contemporary perceptual psychology reviewed here. Furthermore, I will urge that this research presents a fundamental challenge to his entire naturalized epistemology as well as to his empiricist justificatory monism.

There are empirical reasons why Quine should abandon his ban on final causes and intentional entities in light of the empirical support that perceptual psychology has gathered for representations. In particular, if Quine accepts his behavioral test as a vindication for perceptual similarity standards, he must accept, by parity of reasoning, perceptual psychology's behavioral evidence for the claim that bees have objective representations in Burge's sense (which means that they are governed by perceptual norms). Let me quote the passage where Quine delineates the behavioral test to chart an animal's innate perceptual similarity standards (compare above, Sect. 3.1.2):

> Now we can correct our formulation of the behavioral condition for perceptual similarity, to read thus: *a* is shown to be perceptually more similar to *b* than to *c* when the subject has been conditioned to respond in some fashion to all episodes in the receptual neighborhood of *b*, and to withhold that response from all those in the receptual neighborhood of *c*, and is then found to so respond to those in the neighborhood of *a*. (Quine 1974, pp. 17–18)

According to this passage, a subject's perceptual similarity standards can be charted as follows. We condition the subject to respond in a certain way to a certain stimulus; furthermore, we condition the subject *not* to respond thus to another stimulus. Then, we can test whether a third stimulus is perceptually more similar to the first than to the second for the subject as follows. We stimulate the subject in the third way and see whether the subject does or does not react in the manner in which it has been conditioned to react to the first stimulus.

Now compare this procedure to the second of the experiments with bees reconstructed above (compare above, Sect. 4.1, Chittka et al. 2014, p. 439 and Werner 1990). Using Quine's vocabulary, one can say that these bees are conditioned to respond in a certain way to a certain stimulus, namely to fly to the yellow plate whenever presented with one. However, in the test conditions, the bees show that their perceptual similarity standards are keyed to a distal surface property, namely its reflectance: they choose the plate with the same distal property *in spite of* the fact that a different plate caused a proximal stimulus that, in contrast to the chosen plate, matched exactly the proximal stimulus of the training plate.

Given this experimental evidence, it is time for the Quinean to abandon his ban on teleology: the way in which perceptual psychology invokes representations to explain the behavior of bees lives up to the standards that Quine himself maintains to justify his positing of innate perceptual similarity standards.

As a matter of fact, however, Quine decided to ignore perceptual psychology throughout his publishing career. There is no single reference in his monographs to perceptual psychology in Marr's tradition. This does not hold for his posthumously published work: in a passage from his unpublished writings where he sides with Descartes (see below, Chapter 5), he takes evidence from perceptual psychology to establish that we have innate knowledge. The relevant evidence is our ability to represent different perspectives, causing different patterns on the retina, of a given object as the same thing. I think that Quine admits too much: there seems no reason to call this ability of our visual system *knowledge*. Following Burge (2010, p. 541), I would suggest that, since these structured representations are not used, and perhaps not usable as such, to draw inferences, they do not constitute knowledge.

Even if the representations posited by perceptual psychology do not amount to knowledge, Quine's ignoring perceptual psychology may be understandable if it is appreciated what it would mean for Quine's entire naturalized epistemology to take contemporary perceptual psychology seriously—even apart from the fact that perceptual teleology violates Quine's rejection of final causes and hence his eschatological physicalism. As I have shown, contemporary perceptual psychology accepts Quine's doctrine that the proximal stimulus underdetermines the respective distal stimulus. Unfortunately, this is where the agreement between Quine and perceptual psychology ends. Perceptual psychologists presuppose that the visual system overcomes this underdetermination; it postulates that humble creatures such as bees, not to speak of pre-linguistic infants, perceive their distal environment by representing—accurately or inaccurately—particulars with repeatable attributes.

This conception of perception is fundamentally at odds with Quine's insistence that the only information that reaches us through perception is *holophrastic* (compare also Burge 2010, p. 215). First, the structured representations are referential, which means that it is difficult to uphold

Quine's insistence that one cannot credit children with reference until they master pronouns.[9]

More importantly, it is also natural to assume, given such structured perceptions, that fully-fledged, structured statements can be directly falsified by perception. While perceptual representations are no fully-fledged statements, it seems highly unlikely that, in acquiring fully-fledged statements, we do not build on the pre-existing structure given in perceptual representations. This is tantamount to claiming that the evidence human infants rely on to learn their first language is not holophrastic, which overturns the foundations of Quine's naturalized epistemology.[10]

To conclude this section, I consider two possible ways in which a Quinean could argue that the research from perceptual psychology, as rehearsed here, is irrelevant. Both of these ways allow her to fully grant the solidity of the research that establishes that honeybees have objective representations.

First, the Quinean could point out that the perceptual psychological research considered here concerns *honeybees*, and not human pre-linguistic infants; hence, he could declare this research irrelevant for naturalized epistemology. However, there is extensive experimental research that suggests that human infants can achieve color constancy (for an overview on this research, compare Wilcox 1999). The vast majority of this research employs the habituation and preferential looking method (this method is discussed in detail below, Chapter 5). For instance, Yang et al. (2013) used the displays for habituation and test that are shown in Fig. 4.2. The display on the right-hand side, the test display, differs from the habituation display on the left-hand side in a change of illumination: phenomenologically speaking, the former display is under blue light. Furthermore, the right smiley in the display on the right-hand side differs in chromaticity,

[9]For the role of pronouns in Quine's account of the genesis of reference to objects, compare Quine (1974, p. 89). By the same token, perceptual psychology questions Quine's inscrutability of reference: if the visual system, as it were, automatically produces representations that pick out environmental particulars and assign repeatable attributes to them, it is difficult to maintain that it is not determinate, but up to pragmatic decisions from our part, whether we perceive an individual rabbit or a temporal stage of a rabbit process. It is natural to assume that our visual system spares us the pain of deciding between these options. Burge (2010, p. 215) also addresses this issue.

[10]As I will argue below, in Sect. 6.2.1, this also questions the main pillar of Quine's case for empiricist justificatory monism.

Fig. 4.2 The habituation and test displays used by Yang et al. (2013, p. 130) (Reprinted with permission from Elsevier)

that is, in surface color, from the right one on the left-hand side, while the color of the respective left smileys is identical.

The idea behind this set-up is that if infants are able to achieve color constancy, they will find the left smiley of the test display familiar, and hence uninteresting, and therefore focus on the right smiley in the test display. If they do not have color constancy, in contrast, they will find both smileys in the test display equally novel and interesting, as the proximal stimuli caused by both smileys are different from the ones caused by the two smileys in the habituation display, due to a change in illumination. The results do indeed show that infants as young as 4.5 months have a marked looking preference for the smiley on the right-hand side of the test display.

In addition to this experimental difference, there is another reason why Quine should find it unattractive to accept that honeybees as well as adult humans have color constancy, and the corresponding objective representations, but maintain that pre-linguistic human infants do not have it and hence do not rely on the corresponding representations in their acquisition of their first language. Color constancy, as successfully explained by perceptual psychology, is not something that has to be learned. Rather, it is something that our visual system achieves automatically. This means that this ability of the human visual system is independent of the learning of a language—a point that is reinforced by the fact that honeybees, which surely do not have a language, do achieve color constancy as well.

The second possible way how Quine could justify his ignoring objective representations in perception for his naturalized epistemology consists in pointing out that the bee cannot tell you what the respective square in the experiment detailed above looks like, phenomenologically. Our reason for attributing a representation as of yellow (see above, Sect. 4.1) to the honeybee consists entirely in a formation principle employed by perceptual psychology to explain the achievement of color constancy and ultimately the behavior of the bee. However, Quine should be the last philosopher to be troubled by this lack of phenomenology. His notion of stimulus meaning, as detailed above (Sect. 2.2.1) is about as externalist as it gets; hence, he should welcome the way in which perceptual psychology treats of these attributes.

Note, furthermore, that perceptual psychology's attribution of objective representations to honeybees raises perhaps an even more fundamental problem for Quine's naturalized epistemology. Quine's eschatological physicalism consists in the conviction, or rather the hope, that future developments in science will allow for a fully mechanistic explanation of how humans acquire science. However, if it is the case that irreducibly intentional, non-mechanistic entities have to be used already when explaining the behavior of such humble being as honeybees, this very hope of eschatological physicalism is put in serious question: certainly, it is much more difficult to give a mechanistic explanation of the scientist's behavior than it is to give a mechanistic explanation of the bee's behavior.

In short, even by his own experimental standards, Quine should accept that the experiments reviewed here show that honeybees can achieve color constancies and hence have objective perceptual representations in Burge's threefold conception. Once this is granted, it is difficult for Quine to uphold his basic principle that the perceptual evidence for science as well as for first-language acquisition is holophrastic. Furthermore, granting that bees have such objective representations questions the very mechanistic fabric of Quine's naturalized epistemology.[11]

[11] Hanjo Glock worries that I am ignoring an important distinction in my critique of Quine: He holds that it is one thing to claim that intentional concepts such as *representation* cannot be semantically reduced to non-intentional vocabulary, but quite another thing to claim that this semantic fact implies anything with regard to the furniture of nature. While this may be true as a general observation, it is misguided in both Quine's and Burge's case. Quine famously held that

4.4 Conclusion: A Step Forward, but Only a Step

To conclude this chapter, I consider whether perceptual psychology could possibly succeed where Quine has failed: to give a scientific account of the genesis of successful science as well as of its justification. After all, having scientifically accounted for structured, intentional states of animals, including humans, does look like an important step in this direction.

I am inclined to agree to this insofar as it is kept in mind that this only constitutes a first step into the right direction: being able to explain scientifically, in a detailed and empirically significant manner, how primitive representations are produced, does constitute a first step in the direction of scientifically explaining the genesis of science and its justification: the step from there to fully-fledged propositional thought is a smaller one than the one from mere sensory discrimination.

However, I wish to emphasize two aspects of these explanations of perceptual representations that should caution the naturalized epistemologist against being too optimistic. First, this account does not present an explanation how these representations could have developed from non-representational states. This is disappointing for these naturalized epistemologists, like Quine, who want to tell a story how intentional structures evolved from non-intentional ones.

Secondly, and more importantly, the kind of intentionality studied by perceptual psychology may not suit the purposes of naturalized epistemologists. Intentional perceptual representations are produced by the visual system, not by the central cognitive faculty that is typically taken to be responsible for judgments, logical inferences, and thinking in general. It would be implausible to think that honeybees understand that the surface

ontology recapitulates philology (this is the slogan of Quine 1960). This means, *pace* Glock, that semantics does determine ontology. As I have shown above, Burge shares this Quinean position: He holds that the best reason to accept the existence of some sort of thing is that it centrally figures in successful scientific explanation.

reflectance stays the same in spite of looking differently.[12] In contrast, scientific theorizing does seem to involve the central cognitive faculty.

Furthermore, objective representations, as they are achieved by the visual systems of honeybees, are produced by law-like transformation processes that are almost deterministic. This very property explains why these representations can be treated scientifically: their almost-deterministic production means that it can be predicted in impressive empirical detail what kind of representation is produced in what circumstances. Unfortunately, to explain the inherently creative and perhaps unpredictable phenomenon of scientific theory-building, such almost-deterministic processes may be exemplary unsuited.

Still, it seems clear that objective representations in perception, as they are studied by perceptual psychology, do constitute a step towards scientifically elucidating the relationship between scientific theorizing and its evidence. In the next and final chapter of this part of the book, I examine whether empirical science actually has already reached this goal. Maddy's proposal, as I will discuss it in the following chapter, makes exactly this claim.

References

Burge, Tyler. 2010. *Origins of Objectivity.* Oxford: Oxford University Press.

Chittka, Lars, Samia Faruq, Peter Skorupski, and Annette Werner. 2014. Color Constancy in Insects. *Journal for Comparative Physiology A* 200: 435–448.

Freeman Jr., Robert B. 1967. Contrast Interpretation of Brightness Constancy. *Psychological Bulletin* 67 (3): 165–187.

Gilchrist, Alan, Christos Kossyfidis, Tiziano Agostini, Xiaojun Li, Frederick Bonato, Joseph Cataliotti, and Branka Spehar. 1999. An Anchoring Theory of Lightness Perception. *Psychological Review* 106 (4): 795–834.

Marr, David. 2010 [1982]. *Vision.* Cambridge: The MIT Press.

Palmer, Stephen E. 1999. *Vision Science.* Cambridge: The MIT Press.

[12]Thus, Burge (2010, pp. 438–450) emphasizes, in explicit disagreement with Spelke (2004 [1988]), that these representations as of bodies are provided by the visual system, and hence do not need any central cognitive capacity, such as judging, theorizing, or thinking of some sort.

Quine, Willard Van Orman. 1960. *Word and Object.* Cambridge: The MIT Press.
———. 1974. *The Roots of Reference.* LaSalle: Open Court Publishing.
———. 1981. Things and Their Place in Theories. In *Theories and Things*, 1–23. Cambridge: Harvard University Press.
———. 1992. *Pursuit of Truth*, 2nd ed. Cambridge: Harvard University Press.
Richardson, Robert C. 2007. *Evolutionary Psychology as Maladapted Psychology.* Cambridge: The MIT Press.
Spelke, Elizabeth. 2004 [1988]. Where Perceiving Ends and Thinking Begins: The Apprehension of Objects in Infancy. In *Perceptual Development in Infancy.* The Minnesota Symposia on Child Psychology, 20, ed. A. Yonas. Hillsdale: Lawrence Erlbaum.
Warren, William H. 2012. Does This Computational Theory Solve the Right Problem? Marr, Gibson, and the Goal of Vision. *Perception* 41: 1053–1060.
Werner, Anette. 1990. Farbkonstanz bei der Honigbiene, Apis Mellifera. PhD thesis, Free University of Berlin.
Wilcox, Teresa. 1999. Object Individuation: Infants' Use of Shape, Size, Pattern, and Color. *Cognition* 72 (2): 125–166.
Yang, Jiale, So Kanazawa, Masami K. Yamaguchi, and Ichiro Kuriki. 2013. Investigation of Color Constancy in 4.5-Month-Old Infants Under a Strict Control of Luminance Contrast for Individual Participants. *Journal of Experimental Child Psychology* 115 (1): 126–136.

5

Maddy: Conceiving Logic as an Innate Cognitive Mechanism

In this chapter, I examine whether one can take the success of perceptual psychology, as evinced in the preceding chapter, to another level: is it possible to provide a solid experimental account of an innate capacity to fully-fledged propositional thinking, a capacity that is natural insofar as it is not dependent on cultural environments and artifacts like languages?

There is a prominent position in contemporary developmental psychology that essentially makes this claim. I call this position *nativism* (I detail on this conception below, Sect. 5.1). Briefly, nativists claim to provide convincing empirical evidence for two different claims. First, nativists claim that pre-linguistic infants use a sophisticated set of criteria to individuate and refer to instances of a kind of physical object, so-called Spelke objects. Second, nativists claim that pre-linguistic infants have knowledge about the physical behavior of Spelke objects that allows them to judge the possibility or impossibility of certain events involving such Spelke objects. For instance, just like adults, infants know that one solid object cannot easily be penetrated by another. Using a common terminology (see, for instance, Glock 2012, p. 213), nativism claims both that pre-linguistic infants can think *of* certain things, namely Spelke objects, and think *that* such-and-such is the case. The latter is tantamount to attributing beliefs

R. Gubelmann, *A Science-Based Critique of Epistemological Naturalism in Quine's Tradition*, https://doi.org/10.1007/978-3-030-24524-5_5

to the infants in question. Contemporary nativists mostly support their position with results from a research method that is called habituation and preferential looking, which will be discussed in this chapter.

Very surprisingly, the research to be considered here in support of nativism, including the nativist conclusions drawn from it, is mentioned very favorably in Quine's posthumously published work. In a note standing under the title *The Innate Foundational Endowments*, dated March and May 1996, that is, roughly one year after the publication of *From Stimulus to Science*, Quine clearly rejects source-empiricism[1]:

> Descartes thought we had innate knowledge and innate ideas. Locke thought not. I despair of sharpening the issue by defining the term 'idea'. Definition even of 'knowledge' is in trouble since Gettier's challenge of the definition of knowledge as true and warranted belief. However, we need no such sharpening of the issue to see that the evidence favors Descartes over Locke. It is reported that a child in early infancy will show anxiety when glowered at, and will even imitate a frown or smile. The infant will show anxiety also after seeing an object steadily approach a screen, pass behind it, and then fail to emerge in due course at the far side. Further, there is our spontaneous identifying of different perspectives of an angular object as all somehow the same thing. All this is innate, or, as we say, instinctive. Hence innate knowledge, we might say, and no words to express it. (CCE, p. 176)

This is a remarkable passage not only because of its content—Quine accepts that humans have innate knowledge—but also because Quine admits that he has no clear and distinct conception of both knowledge and idea. For the purposes of the present chapter, I cash out innate knowledge of infants in line with my specification of nativism above: pre-linguistic infants can individuate instances of a certain kind of object, namely the so-called Spelke-object. Additionally, they can judge the possibility or impossibility of some basic events involving such Spelke objects.

Furthermore, as I will show in this chapter and have shown in the preceding one, the evidence that Quine cites very cursorily in this passage stems from two different disciplines (putting aside the reports according

[1] I am quoting from the edition by Dagfinn Føllesdal and Douglas B. Quine: *Confessions of a Confirmed Extensionalist and Other Essays*, abbreviated by 'CCE'.

to which infants become anxious when glowered at, and imitate facial expressions). The second piece of evidence that Quine cites is the fact that we are able to realize, without reflection, that different perspectives on an angular object (which cause very different proximal stimuli on the retina) are perspectives on *the same* object. Such phenomena are investigated by perceptual psychology, which has been in the focus above (Chapter 4).[2]

In the present chapter, I delineate (Sect. 5.2) and critically assess (Sect. 5.3) the research method that Quine implicitly alludes to when stating that infants show a marked reaction when they see an object disappear behind a screen but fail to reappear where it would, had it followed a steady trajectory. The method is called habituation and preferential looking, it was pioneered by Elizabeth Spelke in the mid-eighties, and it is currently the main way in which nativists support their position.

Penelope Maddy relies on results from this research method to support one of three main pillars of her own version of naturalized epistemology, what she calls *Second Philosophy of Logic*: she wants to establish, using purely scientific data, (1) that pre-linguistic infants are able to reason according to a specific logical structure, what she calls *rudimentary logic*, (2) that the world itself has this structure, and (3) that humans are so endowed because the world is so structured (compare Maddy 2007, p. 271). I here only examine the first pillar of her program, the claim that pre-linguistic humans reason according to rudimentary logic. This claim directly contradicts Quine's source-empiricism.

Maddy's project, while undermining the empiricist foundations of Quine's specific variety of naturalized epistemology, is clearly intended in the same general spirit (see above, Sect. 1.1.1).

The nativist claims, if warranted, would undermine the basics of Quine's naturalized epistemology: rather than being the end-point of a complicated development that starts with the acquisition of holophrastic observation sentences, beliefs about the physical world would be with infants almost from the start—and certainly before they acquire their first language. In the following Sect. 5.1, I detail nativism, as defended by contemporary

[2] Quine's example would count as a kind of perceptual constancy, namely shape constancy, in contemporary perceptual psychology. In Chapter 4, the focus has been on two other kinds of perceptual constancy, namely lightness and color constancy.

developmental psychologists, the first pillar of Maddy's *Second Philosophy of Logic*, and Maddy's claim that nativism vindicates this first pillar.

5.1 Spelkean Nativism and Maddy's Pre-linguistic Rudimentary Logic

The preferential looking paradigm was pioneered by developmental psychologist Elizabeth Spelke. Since her seminal work in the mid-eighties, she has become one of the most influential proponents of nativism, and arguably one of the most influential living developmental psychologists *tout court*.

Spelke is particularly known for the claim that pre-linguistic infants possess a concept of physical object—what came to be simply called the *Spelke object concept*, which refers to *Spelke objects*. The concept of a Spelke object determines identity criteria for the respective objects, including "spatial contiguity, common fate, and continuous motion" (Maddy 2007, p. 253). According to this specification, the identity of a Spelke object depends on whether it coheres spatially, whether it moves as a unit, and whether it moves without making abrupt changes in its trajectory. Hence, to credit infants with the Spelke object concept means that infants individuate Spelke objects based on these three criteria. For instance, two oranges that are moving in different directions will count as two Spelke objects, since they are neither spatially contiguous, nor moving into the same direction.

Furthermore, infants that possess the concept of a Spelke object have some knowledge, and hence beliefs, about the behavior of Spelke objects. Compare how Baillargeon et al. (1985) conceive of the significance of the so-called drawbridge experiments (which will be discussed below, Sect. 5.2):

> Our experiments provide evidence that by 5 months of age, infants already appreciate two aspects of the behavior of objects. First, they understand that an object continues to exist when occluded, and that it exists not as a disembodied image residing somewhere behind the occluder but as a solid, three-dimensional entity occupying a specific spatial location. Second, they

> understand that an object can move only through space not occupied by other objects. (Baillargeon et al. 1985, p. 206)

According to this passage, pre-linguistic infants, indeed, infants as young as five months, understand that occluded objects continue to exist as solid, physical entities. Furthermore, such infants understand that one object cannot simply move through another one. That is, the infants can appreciate the impossibility of this event. To put it differently, once infants have cognized some x and y as Spelke objects, they believe that x persists, and hence expect that it will not just evaporate when occluded, and they also believe that x and y are solid, and hence understand that x cannot simply pass through y. Naturally, they are surprised if it looks as if x would, impossibly, pass through y.

Maddy is convinced that this nativist position decisively supports one of three core claims of her *Second Philosophy of Logic.*[3] Compare how Maddy states this claim in the following passage[4]:

> […] human beings believe the simple truths of rudimentary logic because their most primitive cognitive mechanisms allow them to detect and represent the KF-structures in the world. (Maddy 2007, p. 245)

I will sketch Maddy's notion of a KF-structure, or, as she later came to call it 'rudimentary logical structure', in a minute. For now, let me focus on the precise claim that Maddy here makes by stating that the most primitive cognitive mechanisms enable us to "detect and represent" these rudimentary logical structures in the world. Maddy (2007, p. 264) makes clear that to be "most primitive" implies, at least, *pre-linguistic.*

Next, it is necessary to get a clear conception of how to read "detect and represent". Such wording suggests a perceptual ability. However, this

---

[3]This is the basic presupposition of her entire case on Maddy (2007, pp. 245–270).

[4]The basic idea behind Maddy's notion of a KF-Structure (an abbreviation for: Kant-Frege-Structure) was to take Kant's notion of logical form and to update it in the way "contemporary students of the subject" would (Maddy 2007, p. 227). In particular, she suggests replacing Kant's subject-predicate relation with Frege's argument-function relation (Maddy 2007, p. 227). On the same page, Maddy proposes an analogous amendment of the Kantian tables of forms of judgment and categories to accommodate Frege's notion of quantification. In later work, especially in Maddy (2014), she has dropped this terminology of KF-structures, and I think it's good riddance: Maddy is in no way committed to any Kantian views on logic, epistemology, or metaphysics.

not only fits ill with her specification of these mechanisms as *cognitive*. On other occasions, she writes that the infant *thinks* according to these rudimentary logical structures, and that the infant *understands* them (see, in particular, Maddy 2014, pp. 100–101). Given her overall project, I submit that Maddy intends the latter, cognitive reading of "detect and represent": She does not merely want to make some claims about the grouping capacities of perceptual systems of infants. Rather, she wants to claim that these infants *think* in a certain way, namely in the form of rudimentary logic.

Maddy (2007, p. 229) specifies her conception of KF-structure, which is synonymous with rudimentary logical structure (see footnote 64). A central point that Maddy makes in this passage is that rudimentary logic is three-valued, since the world contains many objects for which it is not clear whether a given property *P* does or does not apply to them. What I want to emphasize here is less the details of her three-valued logic (which may be interesting in themselves), but rather the general structure of rudimentary logic. This structure includes *objects, properties, relations, and dependencies*. Furthermore, it obviously contains statements, or propositions that can obtain or fail to obtain. Finally, these propositions are of some complexity: they do contain, in addition to conjunctions and disjunctions, universal generalizations and *modus ponens*.

This short sketch of Maddy's concept of rudimentary logical structure is enough to establish that her approach holds the promise of succeeding where both Quine's naturalized epistemology and Burge's perceptual psychology failed, as it were, as attempts at naturalized epistemology (note again that Burge himself never intended his approach as such an attempt): it seems plausible that if Maddy can show that infants do cognize their environment in such rudimentary logical terms, then this can indeed provide a cognitive foundation for science. Furthermore, there is a real-world psychological research program that promises to support Maddy's claim with detailed empirical data.

In the following two sections, I focus on Maddy's case for the claim that human infants think about the world in terms of objects-with-property structures of rudimentary logic (that is, I leave out her analogous cases for relations and dependencies of rudimentary logic). Briefly, this case proceeds as follows: Maddy (2007, pp. 245–270) reviews habituation and preferential looking experiments, accepts the nativist conclusions that Spelke and her fellow psychologists draw from the results of these experiments, and then infers that these conclusions vindicate her claims regarding the logical structure of pre-linguistic infants' thoughts.

This last step may need some further elaboration. Even if it is granted that the respective experiments vindicate nativism, it seems unclear whether nativism as such supports Maddy's central claim about the logical structure of pre-linguistic infants' thinking. Even if it is granted that there is evidence that infants think in terms of Spelke objects with properties, indeed, that they think like adults in some respects, this may not guarantee that they think in terms of the objects and properties of rudimentary logic. This is simply because this is not even clear in the case of full-grown adults. Hence, *a fortiori*, the looking preferences of pre-linguistic infants do not exclude alternative conceptions of logical structure, such as the term-based logic proposed by Oderberg (2005). More needs to be said why the evidence does favor Maddy's specific logical structure at the expense of the numerous alternative proposals.

Note, however, that this worry does not question the basic thrust of Maddy's approach: if it can be empirically shown that pre-linguistic infants do draw inferences about their environments, then this establishes the first pillar of her *Second Philosophy of Logic*—irrespective of the details of the infants' inferential patterns. In the worst case, Maddy might have to make some minor modifications to her concept of rudimentary logic.

In sum, nativism, as it is understood here, claims that pre-linguistic infants can individuate so-called Spelke-objects based on the criteria of cohesion, common motion and continuous motion. Furthermore, pre-linguistic infants can attribute properties such as persistence and solidity to these Spelke objects. Based on this, they are able to understand some aspects of the physical behavior of objects, that is, they think like adults in this regard. Maddy holds, correctly, as I have suggested, that nativism implies the first central claim of her *Second Philosophy of Logic*, according to which pre-linguistic infants think in accordance with rudimentary logic.

5.2 In Support of Nativism: Preferential Looking Experiments

In this section, I detail Maddy's evidence for the claim that pre-linguistic infants reason according to rudimentary logic. More precisely, I focus on the basic part of rudimentary logical structure, objects with properties.

Maddy's argument for the claim that infants cognize the world in terms of Spelke objects with properties is based on the research method called *habituation and preferential looking* (Maddy 2007, p. 247). The basic structure of experiments conducted using this method involves two steps. First, *habituate* infants to a certain scene. This involves repeatedly showing a certain arrangement of items, or a certain event, to infants until they lose interest in the scene or event. It is assumed that infants lose interest if their looking time at the scene or event falls below a certain threshold. For instance, the relevant display could be a yellow ball resting on a blue bar. To habituate the infants to this display, they are placed before a little stage, then the display is presented to them, perhaps by removing a screen, or curtain. As soon as they look away, the display is hidden from them. After that, the display is shown again, etc., until their looking time falls below the threshold.

In the second step of the experiment, the infants are presented with one or several *test displays*. What these test displays show depends entirely on the research interest of the experimenter. To continue the example introduced in the preceding paragraph, if the goal of the experiment is to determine whether infants perceive scenes as novel that only differ in color, the test displays could consist in balls of various colors resting on bars of various colors. If the infants look longer at displays of non-yellow balls resting on non-blue bars than on the habituation display, and given that the form of both balls and bars is the same as in the habituation display, this is taken as evidence that the infants perceive scenes as novel that only differ in color, since they look longer at such scenes. It is then said that the infants prefer (to look at) the display with colors that differ from the habituation colors. Hence, a basic presupposition of the entire method is that infants find novel displays more interesting than familiar ones, and therefore look longer at novel displays.

The key advantage of this method is that it is applicable to infants at a very young age, since it does not require that infants have mastered any sophisticated behavioral abilities such as grasping or crawling: it suffices that the infant is able to focus her gaze on a specific scene.

In the following, I reconstruct the experimental evidence from this preferential looking method that Maddy enlists to show that human infants think according to the structure of rudimentary logical objects with properties. The reconstruction is not exhaustive, but merely representative: I chose two experiments that have been particularly influential in the research domain and on which Maddy puts special weight. Such a representative sample suffices for the purposes of my critical appreciation of the method below (Sect. 5.3).

How Infants Individuate Spelke Objects. First, consider the following experiment that establishes, according to Maddy (2007, p. 251), that infants use continuous motion to determine object identity (as a matter of fact, Maddy (ibid.) never explicitly says so, but she approvingly quotes a passage to this effect). The experiment, or better, the two experiments have been conducted by Spelke et al. (1995, pp. 118–122). In the first experiment, the set-up is as follows: on the stage, there are two screens, one on the left and the other on the right of the stage, behind which rods can move freely. Then, three different groups of four-month-old infants are habituated to three different kinds of events that all take place on this stage (Spelke et al. 1995, pp. 119–120). The first group is shown the continuous bar movement, that is, a bar appears on the very left side of the stage and moves continuously over to the very right side of the stage, passing behind the two screens. The second group is shown the discontinuous bar movement, that is, the bar appears on the left side of the stage, disappears behind the left screen, and then, in due time, another bar appears on the right-hand side of the right screen and moves over to the right edge of the stage. The third group of infants of the same age is habituated to the stage and the setting by showing them events unrelated to the present issue.

Then, all three groups are presented with two different displays of the stage with the screens removed. In the first display, two rods are shown where previously the screens have been located. In the second display, only one rod is shown where previously one of the two screens has been located.

According to the hypothesis of the experimenters, continuous movement constitutes a core criterion that infants rely on when individuating objects; hence, the group that has been shown the continuous movement should be surprised, and hence look longer, at the display of two rods, since they have represented the situation as involving only one rod. In contrast, the group that has been shown the discontinuous movement should be surprised by, and hence look longer at, the display with only one rod, since they have represented the situation as involving two rods. In both cases, the differences in looking time should be significantly above any differences in looking time at the same displays of infants in the control group—otherwise, the differences in looking time could be due to, say, a general preference of infants for more things.

The results of the experiment are as follows. The infants habituated to the continuous movement looked longer at the display of two rods, while the infants habituated to the discontinuous movement looked longer at the display of just one rod. The infants of the control group, finally, looked longer at the presentation of two rods. Unfortunately, the differences between the looking times of these three groups were too small to be significant (Spelke et al. 1995, p. 123). Thus, the interpretation of the outcome of the first experiment is summarized by the experimenters as follows:

> The experiment therefore provides no clear evidence that infants in the continuous condition perceived a determinate number of objects during familiarization, as predicted by the thesis that infants perceive object identity in accord with the continuity principle. (Spelke et al. 1995, p. 123)

To remedy this, the experimenters devised a second experiment, identical to the first one, except for the following aspects: the experimenters used a considerably greater number of infants (47 as compared to 24), the single rod paused when behind the screen so as to match the motions of the discontinuous movement, and the experimenters ensured that the noise produced by both experiments was equivalent (Spelke et al. 1995, p. 124).

In this second study, the infants habituated to the continuously moving rod showed a more significant preference for the two-rod display (i.e. they looked significantly longer at it) than the test group; however, the preference of the discontinually-habituated infants for the one-rod display actually decreased, thus contradicting the predictions of the experimenters.

Thus, Spelke et al. (1995, p. 135) conclude: "Experiments 1 and 2 provide evidence that 4-month-old infants' perception of object identity over occlusion is reliably affected by information on the continuity or discontinuity of object motion." This may be correct, but given that the infants that were habituated to the discontinuous event showed less interest in the display that should be surprising to them, namely the display of just one rod, it seems not entirely clear how exactly this perception is affected by said information.

Infants' Knowledge About the Behavior of Spelke Objects. The most influential kind of habituation and preferential looking experiment that is taken to show that infants understand to some extent what can and what cannot happen to Spelke objects is the so-called drawbridge experiment introduced by Baillargeon et al. (1985) and refined by Baillargeon (1987). Baillargeon (1987) describes her experiment—which is mostly a more sophisticated repetition of the experiment conducted by Baillargeon et al. (1985) with even younger infants (3.5 and 4.5 as opposed to 5 months old, see Baillargeon [1987, p. 656/659])—as follows. Infants are habituated to a screen (the 'drawbridge') rotating back and forth by 180 degrees on a plain. Then, infants are shown four test scenarios. The first two scenarios just show the drawbridge moving 112 degrees and 180 degrees respectively. Scenario three and four show the drawbridge moving towards the box, and then either stopping at the box or apparently going through the box. Scenarios 1–3 are called possible, while scenario 4 is called impossible. The idea is that if the infants look longer at the impossible movement, it is because they understand that it is impossible, as it would otherwise be more familiar to them than the restricted movement. The outcome of these studies is that, overall, infants did indeed look significantly longer at the impossible scenario.

Here is how Maddy comments on a passage from Baillargeon et al. (1985, pp. 195–196):

> Our reasoning was as follows. If infants understood that (1) the box continued to exist, in its same location, after it was occluded by the screen, and (2) the screen could not move through the space occupied by the box, then they should perceive the impossible event to be novel, surprising, or both. (Baillargeon et al. 1985, pp. 195–196)
> [...]
> In fact, the infants looked longer at the impossible event, evidence for their understanding of something like (1) and (2). (Maddy 2007, p. 248)

In this passage, Maddy clearly, even though with some hedging ("something like"), approves of the experimenter's interpretation of the infants' looking preferences: these preferences are explained with recourse to the infants' understanding that it is not possible for the screen to move through the box.[5] Baillargeon (1987, pp. 661–663) addresses a further observation that is of considerable interest (not only because of one of the characters involved): those infants that had to be presented with the habituation display (rotation of the drawbridge of 180 degrees) many times to fulfill the habituation threshold did *not* show a preference for the impossible event—even if the box was replaced by a small figure of Mr. Potato Head. This hints at one of the studies that I will detail below (Sect. 5.3.2).

To sum up, in this section, I have considered a representative—but not exhaustive—sample of the preferential looking experiments that Maddy enlists to support her claim that pre-linguistic infants reason according to rudimentary logical structure. The first experiment is taken to show that pre-linguistic infants individuate objects based on the criterion of common motion: objects that move continuously, even when occluded, are represented over time and space as one persisting object. The second

[5] What Maddy quotes is not the summary of the results of the more sophisticated experiments conducted by Baillargeon (1987), but rather the research hypothesis of Baillargeon et al. (1985). However, the summary given by Baillargeon (1987, p. 663) of her results is essentially equivalent to the research hypothesis of Baillargeon et al. (1985) quoted by Maddy: Baillargeon maintains that her results indicate that infants attribute permanence and solidity, or impenetrability, to Spelke objects, which allows them to suppose that objects cannot move through other objects.

experiment is taken to show that pre-linguistic infants understand that objects persist when occluded and that they are not easily penetrable.

5.3 Preferential Looking Experiments Do Not Vindicate Nativism

In this section, I critically assess whether preferential looking experiments do indeed justify crediting pre-linguistic infants with the sophisticated cognitive abilities that nativists want to attribute to them. I first (Sect. 5.3.1) highlight the high degree of sophistication of these cognitive capacities. After pointing out an empirical problem regarding attributing such sophisticated cognitive abilities to infants, I suggest that doing so needs to be supported by very strong and specific empirical evidence. This latter suggestion is based on a methodological principle introduced by Glock (2013). Then (Sect. 5.3.2), I show that the preferential looking experiments pioneered by Spelke are subject to strong experimental-methodical critique, critique that suggests that non-cognitive explanations[6] rather than cognitive ones can explain the looking behavior of infants in such experiments. Finally (Sect. 5.3.3), I conclude by suggesting that the evidence that can be provided by habituation and preferential looking experiments might be too unspecific in principle to warrant attribution of such high cognitive abilities to pre-linguistic infants.

5.3.1 Nativism Needs Very Specific Evidence

A common understanding of 'understanding', 'believing', 'expecting', etc. requires to be able to distinguish between what one understands, believes, and expects, and what actually is the case: somebody who believes that *p* must understand that there is a reality that may contradict her belief. Similarly, to expect, and to be surprised, the infant needs to be able to think something like the following: *I* have expected such-and-such to occur,

[6]That is, explanations that do not credit infants with any cognitive abilities, but rather explain their behavior with other properties of their psychology, such as simple preferences for more motion or novelty.

but what *actually* occurs is very different. Hence, given the fully intentional interpretations of the infants' behavior in the preferential looking experiments, nativists like Spelke are committed to attribute to the infant the ability to make a subjective-objective distinction: the infant needs to be aware that her expectations, her concepts are hers, and that they may differ from what actually is the case. This ability is usually called epistemic self-consciousness.

A problem with attributing such an epistemic self-consciousness to infants is that it contradicts empirical research. Empirical work shows that infants do not develop the ability to distinguish between what they think is the case and what actually is the case, or simply the capacity to think that they may be wrong, until well after their acquisition of their first language. According to the current developmental accounts of self-consciousness, infants do not develop even a psychological—let alone an epistemic—self-consciousness until about 10 months of age (compare the overview on Bermúdez [1998, pp. 271–275], with empirical details on the development of psychological self-consciousness in infants on Bermúdez [1998, pp. 229–265]). Bermúdez (1998, pp. 69–71) argues against the attribution of concepts to infants along lines that are broadly analogous to the one suggested here.

Can nativists respond to this worry by delineating a different understanding of understanding, expecting, being surprised, etc. that does not require epistemic self-consciousness, and that is much less cognitively demanding in general? Such notions are proposed by Glock (2013, pp. 214–216) and Burge (2010, pp. 267–276). Perhaps they could, but this is exactly not the choice of nativists. Spelke explicitly embraces the sophisticated cognitive structures that have to be in place for an infant to possess the Spelke object concept, and to understand that some things can and others cannot happen to such objects:

> The object concept, on this view, arises from an intuitive theory of the physical world and its behavior. It is present, albeit in impoverished form, near the beginning of life. (Spelke, 2004, pp. 333–334)

Spelke is not very explicit about what exactly she means by an "intuitive theory of the physical world and its behavior", but it is evident that she is willing to bite the bullet and attribute not only a single object concept, but much more substantial cognitive structures to the infant. It is plausible that the notion of a theory of the physical world can be specified in such a way as to support the claim that the infant understands much of the physical behavior of objects and is indeed surprised if objects appear to violate this physical theory.

Note that it is not clear whether being able to individuate Spelke objects based on the three criteria mentioned can be separated from being able to understand some of the physical events that can and that cannot happen to these objects. This means that, for the present case, it may be that thinking of a certain instance of a Spelke object implies being able to think that certain things can or cannot happen to them. For instance, consider the nativist interpretation of the infants' looking behavior in the rod experiment. Infants habituated to the continuous movement looked longer at the display of two rods because they had represented their habituation event as involving one rod only. In contrast, infants habituated to the discontinuous movement should have looked longer at the one-rod display (which they actually did not), because they had represented the habituation event as involving two rods. Can this explanation be separated from the following: infants understand that it is possible that one rod continuously moves across the entire stage, while they understand that it is not possible that one rod discontinuously moves across the entire stage?

Finally, note that the nativists' reasoning takes the general form of an inference to the best explanation: there is a phenomenon to be explained, namely the looking preferences of pre-linguistic infants in preferential looking experiments. Nativists claim that attributing the ability to refer to Spelke objects, and a partial understanding of the things that can and cannot happen to these objects is the best, or the only explanation of this phenomenon (compare Sect. 8.1 for a discussion of abduction as a form of inference).

As such, nativism would almost certainly be a target for the methodological principle called Morgan's canon, according to which, for any behavioral pattern of a given animal, we should always adopt the explanation that attributes the least complex cognitive abilities to the respective animal (compare Glock 2013, p. 147).

Against Morgan's canon, Glock (2013, pp. 147–149) rightly maintains that there are other relevant explanatory virtues than the simplicity of cognitive abilities attributed. For instance, to explain a given behavioral pattern, attributing relatively simple cognitive capacities can be inferior to an explanation that attributes more complex cognitive capacities if the simpler capacities can only explain a very limited range of behavioral patterns—perhaps only the single behavioral pattern at issue. More generally speaking, explanatory unification and generality should be accepted as explanatory virtues along with the simplicity of the cognitive capacities attributed.

Hence, if it would turn out that the nativist explanation of the looking preferences of infants in the respective experiments can only be avoided by positing simpler cognitive capacities that are highly ad hoc, this would not threaten the status of the nativist explanation as the best explanation for the infants' behavior. I take this to be the essence of Glock's more modest alternative to Morgan's canon, called *Glock's handgun* (Glock 2013, p. 148).

However, even when aimed at merely with Glock's handgun, and not with Morgan's canon, it takes very substantial evidence for the nativist explanation to hold its ground: claiming that pre-linguistic infants are able to think like adults about parts of the physical world involves attributing to these infants about as sophisticated cognitive abilities as possible. Furthermore, according to Glock's handgun, the simplicity of the cognitive capacities used is not irrelevant—it is just not the only explanatory virtue that counts. Hence, while such an explanation is very likely to be pulverized by Morgan's canon, it may survive Glock's handgun if it offers a high degree of unification and generality, and if it is extremely well-supported by the empirical evidence. In the following section, I argue that habituation and preferential looking experiments cannot provide such evidence.

5.3.2 The Experiments Do Not Provide Sufficiently Specific Evidence

In this section, I critically assess the evidence that preferential looking experiments can provide for nativism. In the recent empirical literature, it was the interpretation of the infants' looking time in the drawbridge scenario that has received the greatest amount of critical consideration.

Rivera et al. (1999) provide an elaborate and empirically supported alternative to the nativist explanation for the looking behavior of infants in the drawbridge experiments. Remember that nativists infer identification of Spelke objects and understanding of physical possibility and impossibility from the fact that infants look longer at the so-called impossible event. This is the scenario where the drawbridge moves a full 180 degrees and thereby apparently, and impossibly, passes right through a box in its way.

Rivera et al. (1999, p. 433) hypothesized that the longer looking time may not be due to the infants' understanding that this event is impossible, but simply to their preference for more motion. Of course, Baillargeon (1987) tried to forestall this explanation by testing the looking time of a so-called control group of infants of the same age that did receive the same habituation, namely the repeated display of the "drawbridge" performing a 180 degrees movement, see Baillargeon (1987, pp. 657–659). Then, the control group was shown merely two different movements of the "drawbridge", one of them 180 degrees, the other 112 degrees, without any boxes. It turned out that the infants in the control group did not look longer at the 180 degrees movement, which Baillargeon (1987, p. 659) took as evidence that it is not simply the infants' preference for more motion that accounts for their looking longer at the impossible event.

Rivera et al. (1999) were aware of this control group; however, they suspected that what made the infants in the control group look for roughly the same amount of time at both movements is that the *novelty* of the shorter movement—which was not shown to the infants during habituation—cancels out the effect that *more motion* would otherwise have on the infants' looking time. Hence, these researchers suspected that it was an artifact of the habituation phase, causing infants to find the 180 degrees movement familiar and therefore less interesting, that infants did not display their preference for more motion in the control group.

Furthermore, Rivera et al. (1999) hypothesized, if it was indeed the case that the infants understand that objects are impenetrable, then there would be no habituation needed for these infants to prefer the impossible event. Therefore, the research group conducted a first experiment, which is analogous to the one by Baillargeon (1987), just without a control group and, importantly, without any habituation phase. It turned out that even without habituating, the infants looked longer at the so-called impossible scenario, where the drawbridge appears to go through the box, see Rivera et al. (1999, pp. 430–431).

To confirm the suspicion that it was a combination of novelty and more motion that explains the infants' looking time rather than their finding one event impossible, the researchers conducted a second experiment, which is analogous to what Baillargeon (1987) conducted as the control condition experiment, again with the important difference that there is no habituation phase. In this experiment, infants were simply shown two displays, one of them showing the drawbridge moving 112 degrees, the other showing the drawbridge moving 180 degrees. Put briefly, Rivera et al. (1999, pp. 432–433) report that the infants looked reliably longer at the 180 degrees motion than at the 112 degrees motion—without there being any box in the way. Hence, infants not previously habituated to any movements found both the possible and the impossible 180 degrees movement more interesting than the 112 degrees movement.

Based on this finding, Rivera et al. (1999) conclude that what explains the infants' looking time even in the case of the so-called impossible display is a simple preference for more motion. Furthermore, they suggest that the results of the habituation experiment conducted by Baillargeon (1987) actually reinforces their point: as the 112 degrees movement was novel to the control group, it is remarkable that they did not show a pronounced preference for it; their finding more motion more interesting can explain this absence of a novelty-effect.

In addition to this critique of the drawbridge experiments, Parke and Gauvain (2002, pp. 281–284), introducing developmental psychology, cite research by Bogartz et al. (2000) that also casts doubts on the implications that nativists like Spelke draw from the drawbridge experiments. In this study, the researchers wanted to corroborate their hypothesis that familiarity and unfamiliarity can account for the prolonged looking times in habituation and preferential looking experiments in general, and in the drawbridge experiments from Baillargeon et al. (1985) and Baillargeon (1987) in particular.

To verify their hypothesis, they conducted two experiments. The main methodological difference to Baillargeon et al. (1985) and Baillargeon (1987) is that they split up the infants into three groups, each of which was then habituated to a different event (each event being shown three times during habituation phase, see Bogartz et al. [2000, p. 410]): one to the drawbridge moving 180 degrees without obstacle, one to the drawbridge moving 180 degrees and apparently going through the obstacle, and one to the drawbridge moving 120 degrees and stopping at the obstacle. Then, each group was tested by being shown each of the three displays four times.

Before considering the results, a remark on effects of habituation: the looking time, during habituation, of those infants habituated to the impossible event did not show any significant difference to the looking time of the infants habituated to the two possible events (Bogartz et al. 2000, p. 411). This should give pause to those who think that the main variable influencing infants' looking times is what they conceive, based on the properties they attribute to objects, as possible or impossible events. If infants really thought that objects are impenetrable, and hence, that it is impossible for one object to move through another, the infants that were shown the impossible event should have looked longer at it already during habituation.

Note, furthermore, that the same study also brings bad news for non-cognitivist explanations of the drawbridge-experiment. Bogartz et al. (2000, p. 411) note that they could not confirm (Rivera et al. 1999), who claim that infants simply find more motion more interesting. Based on the latter study, one would expect the infants in the former study to look reliably longer at those displays with more motion.

The results of the test trials, found on Bogartz et al. (2000, p. 412), are as follows: in short, infants simply looked longer at those events that were new to them, given their habituation; those habituated to the 180 degrees movement—with or without box—found the possible event much more interesting than the impossible event. The only group that found the impossible event more interesting is the one that had been habituated to the 120 degrees movement with an obstacle.

In a second experiment, Bogartz et al. (2000) further examined the effects of familiarity for the outcome of the drawbridge experiment. This time, they hypothesized that infants are not, as it were, content with being shown a display 3 times—they want to see it more often. The experiment proceeded as follows: They divided the infants up in four groups, which were each habituated to either the 120 or the 180 degrees movement and habituated 3 or 7 times respectively. So, during habituation, group 1 saw the 120 degrees movement 3 times, group 2 saw the 120 degrees movement 7 times, group 3 saw the 180 degrees movement 3 times, group 4 saw the 180 degrees movement 7 times. The four test displays then consisted of 180 or 120 degrees movements, with or without obstacle. In short, the results suggest that the 3 habituation events increased, rather than decreased, the respective infants' interest in the same displays, while seven habituation displays clearly decreased their interest in the respective displays. Again, whether the displays were possible or impossible from an adult's point of view was entirely insignificant (Bogartz et al. 2000, pp. 421–422).[7]

It is true that this is just one of several experiments that nativists enlist to support their case. However, the drawbridge experiments are by far the most famous instances of habituation and preferential looking experiments. Furthermore, the critiques of Rivera et al. (1999) and Bogartz et al. (2000) focus on the impact of habituation on the results of the drawbridge experiments. As, by definition, all the habituation and preferential looking experiments include habituation, it is likely that versions of their critiques apply to other preferential looking experiments as well.

For instance, consider the rod experiments discussed above (Sect. 5.2), where the behavior of the infants habituated to the discontinuous movement did not fit the nativist explanation: they showed less interest in the one-rod display than they should, given that they individuate objects based on continuous motion and hence should have represented the situation as involving two, not one rod. Perhaps, given the habituation, which involves two screens on the stage, the two-rod display presented infants with just the right mixture of novelty and familiarity to capture their attention (this is at least suggested by Bogartz et al. (2000), who found that, to a certain point, infants prefer displays that they have seen already)? Or perhaps infants have a basic preference for 'more things' on the stage, which caused them to look longer at the display of two rods than at the one of just one rod (which is supported by the behavior of the infants in the control group); however, they also prefer novel displays, which implies a looking preference for the display showing just one rod, and which, combined

[7] Baillargeon et al. (1985) and Baillargeon (1987) report to have shown the infants the habituation display 6–9 times, depending on the individual decrease in looking time.

with the preference for more things, yields the looking behavior observed in the experiment.

Finally, note that, according to Tyler Burge's philosophy of perceptual psychology, the behavior of the infants in these experiments should be explained by attributing to the visual system of the respective infants the capacity to produce certain perceptual constancies (Burge 2010, pp. 438–450). In contrast, nativists explicitly favor a cognitive explanation of the looking times of infants.[8]

These three alternative explanations of the infants' looking behavior in the rod experiment are all *prima facie* solid explanations with at least some empirical support. They are not ad hoc, and they allow for at least the same level of unification and generality as the nativist explanation. Hence, none of them is threatened by Glock's handgun. None of them, however, dovetails with the nativist's claims—and it seems very difficult to conceive of experimental set-ups using the preferential looking method that can eliminate these, and all the other conceivable alternative explanations. This is because the evidence that can possibly be provided by preferential looking experiments—the looking times of infants—seems too unspecific to warrant the strong conclusions that nativists want to draw from them.

This critical assessment of the preferential looking method is reinforced by the so-called *paradox of early permanence*: as Meltzoff and Moore (1998, pp. 202–203) detail, the paradox consists in inconsistent empirical findings. On the one hand, preferential looking experiments are taken to show that 3-month old infants "understand the continued existence and movements of occluded objects" (Meltzoff and Moore 1998, p. 202). On the other hand, infants do not recover hidden objects, that is, they do not search for them, until about 8 months of age, suggesting that they do not understand that hidden objects persist until that age. It certainly supports critics of nativist interpretations of preferential looking experiments that the results obtained through such experiments are in tension with the results obtained by other experimental methods. Even more, these latter results are based on much more specific and sophisticated behavior than mere looking time.

[8]Another proposal to explain the behavior of infants in the rod experiment is due to Hatfield (2009, pp. 241–248). Hatfield only attributes reference to a bounded trackable volume, and not to instances of the kind Spelke object to infants.

5.3.3 Conclusion: Insufficient Evidence for Innate Logic

In this chapter, I have examined one of the three main pillars of Maddy's *Second Philosophy of Logic*: the claim that pre-linguistic human infants can reason in accordance with rudimentary logic. To substantiate her claim, Maddy almost exclusively relies on nativist interpretations of the behavior of infants in so-called habituation and preferential looking experiments. According to these nativist interpretations, the looking preferences of pre-linguistic infants in such experiments are best explained by crediting them with the ability to refer to so-called Spelke objects, and with a partial understanding of what can and what cannot happen to these objects.

Nativism, so conceived, directly contradicts Quine's source-empiricism. Hence, if well-supported, it would undermine virtually every part of Quine's naturalized epistemology. In contrast, it seems clear that nativism decisively supports the main pillar of Maddy's *Second Philosophy of Logic*, according to which humans are innately disposed to pick up and represent rudimentary logical structures.

I have reconstructed a representative sample of the experiments that Maddy invokes in support of her claim. The first experiment aims to show that infants use the criterion of continuous motion to individuate Spelke objects. It aims to show this based on the looking times of infants at displays of one or two rods, after having been habituated to continuous and discontinuous rod movements respectively. The second kind of experiment, the so-called drawbridge experiment, is taken to show that infants understand that occluded objects persist and are not penetrable. This is inferred from the experimental datum that they are looking longer at the so-called impossible event, where the drawbridge appears to penetrate a box, or Mr. Potato head, than at various so-called possible events.

Finally, I have critically assessed these experiments on two levels. First, I have outlined the high degree of sophistication of the cognitive abilities that nativists want to attribute to infants to explain their behavior in preferential looking experiments. I have suggested that being able to understand, to expect, and to be surprised in the nativist understanding of these terms requires an epistemic self-consciousness. However, empirical research suggests that pre-linguistic infants do not have such a consciousness. More generally speaking, I have pointed out that the nativists' inference is abductive in form: it infers nativism as the best explanation of the infant's looking preferences in the experiments. Based on the methodological principle called Glock's handgun, I have urged that it takes much in terms of empirical evidence to support this abductive inference.

Second, I have discussed empirical studies that imply that the habituation and preferential looking method probably cannot provide such empirical evidence. I have focused on the drawbridge experiments, as conducted by Baillargeon et al. (1985) and Baillargeon (1987). I have argued that there are strong reasons, arising simply from paying close attention to the experimental set-up, against taking the results of these experiments to support nativism. First, Rivera et al. (1999) provide evidence that a combination of simple preferences of the infants, namely a preference for novelty and more motion, can explain the experimental results, once the side-effects of habituation (infants find the 180 degrees movement familiar and hence less interesting) are taken into account. Bogartz et al. (2000) also focus on unintended effects of the habituation phase of the drawbridge experiments. They show that the most reliable way to predict the looking time of infants is simply to determine which event is new to them, given the habituation that they have been subjected to. In analogy to Rivera et al. (1999), they show that no cognitive explanation is needed to explain the infants' looking times in these drawbridge experiments (let alone a cognitive explanation of the nativist's caliber). Rather, simple preferences for novelty and more motion suffice, once the side-effects of habituation are taken into account.

I have then urged that these observations question the nativist's claim that her explanation is the best one for the looking preferences of infants in the respective experiments. There are good, non-ad-hoc explanations of the infants' looking preferences for both the rod and the drawbridge experiment that do not attribute cognitive capacities of the nativist's caliber to the infants. Hence, it seems questionable why we should accept the nativist explanation. This holds even if we are not aiming Morgan's canon, but merely Glock's handgun at this nativist explanation. Furthermore, I have urged that it lies in the nature of habituation and preferential looking experiments that such alternative explanations are always possible.

This problem is reinforced by the fact that these nativist explanations themselves stand in tension with other empirical insights, such as the development of epistemic self-consciousness and the recalcitrant data from experiments based on search behavior of infants (leading to the so-called paradox of early permanence, see above, Sect. 5.3.2).

When it comes to these recalcitrant data, Maddy (2007, p. 300) only refers to Huntley-Fenner et al. (2002), who reference empirical work that questions nativist claims, but who are themselves supportive of the nativist interpretation. Hence, I cannot take into account her response to the worries expressed in this chapter.

I therefore conclude that habituation and preferential looking experiments are the wrong kind of experiments to support nativism: the gap between the shallow data—looking preferences of infants—and the highly specific claims that nativists want to support with these data is too broad. (Note that I am not claiming that the dispute between nativists like Spelke and her critics, such as Rivera, has been decided in favor of the opponents of nativism. Rather, I hold that it is undecided, and as I shall suggest in the following, perhaps undecidable on the basis of habituation and preferential looking experiments.)

This verdict is fallible: it could be that researchers become so astute at devising new setups that the heavily cognitive explanation favored by nativists for the looking preferences of infants becomes the only possible one. For instance, it could be that experiments along the line of the rod experiment discussed above (Sect. 5.2) establish beyond doubt that infants represent one single rod in the continuous display, but two rods in the discontinuous display. This requires, for instance, that the experimental data can exclude the explanation that Burge (2010, pp. 438–450) urges, according to which the respective infants' visual systems are capable of achieving a specific kind of perceptual constancy. Similarly, it is not entirely impossible that researchers develop experiments in the drawbridge-tradition that leave as the only acceptable explanation for the infants' looking times that they understand what is physically possible, and what is not physically possible.

However, as it were, I would not hold my breath until such evidence is available. It is just terribly difficult to develop experimental set-ups such that the looking time of infants in these experiments leaves only one explanation, namely that infants think like adults about parts of the physical world. The looking time of infants seems too unspecific a behavior to achieve that. Alternative, much simpler explanations such as preferences for novelty or more motion appear impossible to exclude with this experimental method.

This is of course not to say that the habituation and preferential looking method as such is useless. It seems natural, and broadly in the spirit of behavioral tests specified by Quine himself (compare the procedure that Quine suggests to chart an animal's innate similarity spaces, Sect. 2.1.3, p. 37) that these experiments can serve to elucidate the standards based on which infants perceive events as similar and dissimilar. The doubts expressed in this chapter only concern the attempt to support nativist claims with these habituation and preferential looking experiments.

In particular, these doubts do not concern the experiments discussed above (Sect. 4.1) in the context of perceptual psychology. This is because there exists a twofold contrast between the experiments relied on by Burge and those relied on by nativists. First, in contrast to innate logic and theories of the world, lightness constancy is a simple feature, straightforwardly observable in behavior: if the bee returns to the yellow square with the reward although another square is more like the training plate in terms of the proximal stimulus it causes on the bee, this directly justifies attributing lightness constancy to the bee. In contrast, as detailed above, the longer looking time of an infant at a so-called impossible display can have, in addition to Spelke's and Maddy's favored nativist explanation, a number of alternative causes.

Second, the behavior of bees—flying to the reward—is much richer than the behavior of the infants—looking longer at certain displays. Thanks to this, the experiments can be set-up so as to provide much more specific evidence than mere looking preferences can possibly provide. In particular, in experiments such as the one conducted by Werner (1990), one can explicitly present the bees with a choice between differently composed proximal stimuli, and observe how they behave.

In short, the examinations of this chapter suggest that the nativist claims are insufficiently established as long as they are not supported by more solid evidence than the one currently, and probably possibly, provided by habituation and preferential looking experiments with infants. This means that the attempt to take the next and final step of naturalized epistemology has failed so far (and perhaps permanently): while it has been possible to give an austerely scientific account of proto-propositional representations in perception, the attempt to do the same for *bona fide* propositions has failed.

5.4 Conclusion: Dim Prospects for Naturalized Epistemology

Let me briefly rehearse the results of my discussion of naturalized epistemology, which constitutes the first part of this study. In Chapter 2, I have detailed Quine's account of the infant's first steps into language, that is, I have examined how she acquires her first observation sentences. After introducing the physicalistic and empiricist background of this account, I have reconstructed Quine's elegantly physicalistic account of these first steps in *Word and Object*, and I have detailed a basic flaw in this account.

Then, in Chapter 3, I have traced in detail Quine's attempt to repair this basic flaw in this account, namely the fact that stimulus meaning, as he conceives it, is private, that it varies from individual to individual. I have argued that this leaves him without an explanation how communication is so much as possible. Quine's final solution to this problem is a preestablished harmony, achieved through natural selection, of innate perceptual similarity standards. Thanks to these standards, two humans experience the same situations—which lead, due to non-homology, to different patterns of sensory stimulations of the two humans—as relevantly similar. This furnishes the non-verbal reference points that the infant needs to acquire her first language. Furthermore, it makes agreement of competent speakers, including scientists, more than merely verbal agreement, as their verbal behavior is anchored in a shared way of perceiving the world: thanks to the preestablished harmony, any two competent speakers of a given language will assent to a given (holophrastic) observation sentence in exactly the same situations, because they unanimously perceive them as similar in the relevant sense.

Also in Chapter 3, I have developed two immanent critiques of Quine's mature account. First, I have argued that Quine's recourse to natural selection to explain why our innate perceptual similarity standards stand in a preestablished harmony and why they run with the grain of nature (Quine's deliberately vague phrasing) probably cannot be cashed out in fully-fledged scientific terms. This leaves him with no scientific explanation of why communication is possible and why science is successful. Furthermore, I have argued that Quine's account how the infant acquires her first observation sentences tacitly attributes to the infant the ability to conceive of the utterances of her parent as what they are, to wit, speech acts that are intended to fulfill a certain communicative purpose. Furthermore, I have urged that, once Quine's account is purged of these communicative purposes, it is very implausible that the infant acquires even a handful of observation sentences.

In Chapter 4, following the lead of Burge (2010), I have introduced contemporary perceptual psychology. Then, I have argued that Burge is right that this discipline irreducibly posits objective representations in his threefold understanding. I have also urged that there are good empirical reasons why Quine should take this discipline seriously, and hence accept its central ontological posit. However, I have also delineated the fundamental challenge that accepting objective representations, thus understood, constitutes to Quine's naturalized epistemology. They are not only diametrically opposed to Quine's eschatological physicalism. Since they are structured in a particular and a general, repeatable part, they also contradict his holophrastic conception of evidence. Finally, I have outlined that Burge's position does constitute a first step into the naturalized epistemologists' direction: According to Burge, there is a scientific account, which is empirically supported and spelled out in mathematical detail, how structured representations in perception are achieved by the visual systems of many mammal species. While not being a scientific account of predication properly so-called, it is an account of proto-predication.

In this chapter, I have assessed whether this first step can be complemented by another, even more momentous one by detailing the empirical evidence that Maddy musters for the first central claim of her *Second Philosophy of Logic*. According to this claim, prelinguistic human infants are able to reason according to rudimentary logic. This claim directly contradicts Quine's source-empiricism. I have suggested that the evidence that Maddy cites for this claim is, at least for now, but probably in principle, unconvincing: the habituation and preferential looking experiments provide too unspecific data to vindicate Maddy's claim. The evidence from these experiments is also too unspecific to vindicate Spelkean nativism. Hence, I have not found reason to assume that naturalized epistemology is successful. What remains is the first step identified in Burge's account.

To conclude, let me point out a basic tension in the very project of naturalized epistemology. This tension is certainly not shockingly new, but I have underpinned it with substantial empirical evidence in the preceding chapters: it seems that one can either get a proper scientific theory supported by strong empirical evidence, or one can get a theory that purports to explain proper human cognition (and predication in particular), but not both. Burge's account falls in the first category: his reading of perceptual psychology is well-supported by empirical evidence, but it fails to be a theory of human predication properly conceived (which is also not something that Burge intends his theory to achieve).

Maddy's account falls in the second category: being able to reason according to rudimentary logic certainly involves the central cognitive faculty of humans, the same faculty that is in play in scientific theorizing. Hence, while Maddy is aiming high with her nativism in terms of the cognitive capacities she credits pre-linguistic infants with, this nativism would plausibly really explain why humans are able to successfully pursue science. However, it is an intrinsic problem of her high goal that it may not be achievable with the humble means of habituation and preferential looking—and perhaps not with any empirical method that does not involve linguistic communication to determine whether the respective being reasons according to rudimentary logic. This tension is one that any naturalized epistemology will face, and one that presents these attempts with a formidable challenge.

References

Baillargeon, Renée. 1987. Object Permanence in 3 1/2- and 4 1/2-Month-Old Infants. *Developmental Psychology* 23 (5): 655–664.

Baillargeon, Renée, Elizabeth Spelke, and Stanley Wasserman. 1985. Object Permanence in Five-Month-Old Infants. *Cognition* 20: 191–208.

Bermúdez, José. 1998. *The Paradox of Self-Consciousness*. Cambridge: The MIT Press.

Bogartz, Richard S., Jeanne L. Shinskey, and Thomas H. Schilling. 2000. Object Permanence in Five-and-a-Half-Month-Old Infants? *Infancy* 1 (4): 403–428.

Burge, Tyler. 2010. *Origins of Objectivity*. Oxford: Oxford University Press.

Glock, Hans-Johann. 2012. Thought, Judgment and Perception. *Grazer Philosophische Studien* 86: 207–221.

———. 2013. Animal Minds: Philosophical and Scientific Aspects. In *A Wittgensteinian Perspective on the Use of Conceptual Analysis in Psychology*, 130–152. Basingstoke: Palgrave Macmillan.

Hatfield, Gary. 2009. Getting Objects for Free (or Not): The Philosophy and Psychology of Object Perception. In *Perception and Cognition: Essays in the Philosophy of Psychology*, 212–255. Oxford: Clarendon Press.

Huntley-Fenner, Gavin, Susan Carey, and Andrea Solimando. 2002. Objects Are Individuals but Stuff Doesn't Count: Perceived Rigidity and Cohesiveness Influence Infants' Representations of Small Groups of Discrete Entities. *Cognition* 85: 203–221.

Maddy, Penelope. 2007. *Second Philosophy. A Naturalistic Method.* Oxford: Oxford University Press.

———. 2014. A Second Philosophy of Logic. In *The Metaphysics of Logic*, ed. Penelope Rush, 93–109. Cambridge: Cambridge University Press.

Meltzoff, Andrew N., and Keith M. Moore. 1998. Object Representation, Identity, and the Paradox of Early Permanence: Steps Toward a New Framework. *Infant Behavior and Development* 21 (2): 201–235.

Oderberg, David S. 2005. Predicate Logic and Bare Particulars. In *The Old New Logic*, ed. David S. Oderberg, 183–210. Cambridge: The MIT Press.

Parke, Ross D., and Mary Gauvain. 2002. *Child Psychology. A Contemporary Viewpoint.* New York: McGraw-Hill.

———. 2008. *Confessions of a Confirmed Extensionalist and Other Essays*, ed. Dagfinn Føllesdal and Douglas B. Quine. Cambridge: Harvard University Press.

Rivera, Susan M., Ann Wakeley, and Jonas Langer. 1999. The Drawbridge Phenomenon: Representational Reasoning or Perceptual Preference? *Developmental Psychology* 35 (2): 427–435.

Spelke, Elizabeth. 2004 [1988]. Where Perceiving Ends and Thinking Begins: The Apprehension of Objects in Infancy. In *Perceptual Development in Infancy*. The Minnesota Symposia on Child Psychology, 20, ed. A. Yonas. Hillsdale: Lawrence Erlbaum.

Spelke, Elizabeth, Roberta Kestenbaum, Daniel J. Simons, and Debra Wein. 1995. Spaciotemporal Continuity, Smoothness of Motion and Object Identity in Infancy. *The British Journal of Developmental Psychology* 13: 113–142.

Werner, Anette. 1990. Farbkonstanz bei der Honigbiene, Apis Mellifera. PhD thesis, Free University of Berlin.

6

Quine's Empiricist Justificatory Monism

In the preceding part of my study (Chapters 2 to 5), I have examined naturalized epistemology. The present part (Chapters 6 to 9) critically assesses empiricist justificatory monism. Justificatory monism is here understood as defined above (Sect. 1.1.1): justificatory monists maintain that there is only one fundamental kind of justification, which implies that any statement whatsoever that is necessary for science can be justified by this one basic sort of justification.

So far, this explication of justificatory monism does not involve any restrictions regarding the precise explication of scientific justification. Hence, even a thoroughly rationalist conception, according to which any claim is scientifically justified that can be perceived (with the eye of reason) clearly and distinctly could constitute a justificatory monism.

For the purposes of this study, however, it is expedient to focus on empiricist justificatory monisms, that is, on justificatory monisms whose explication of justification is recognizably empirical. This is not only clearly in Quine's spirit; it is also a natural restriction because contemporary naturalists are no hard-core rationalists; rather, they emphasize the importance of empirical evidence for the justification of scientific as well as traditionally epistemological claims.

R. Gubelmann, *A Science-Based Critique of Epistemological Naturalism in Quine's Tradition*, https://doi.org/10.1007/978-3-030-24524-5_6

The general claim that I defend in this second part is that empiricist justificatory monism cannot be defended. The basic reason for this is that for each given empiricist justificatory monism, its respective explication of empirical-scientific justification cannot be justified in the way it itself specifies. This means that empirical science, on any given explication of empirical-scientific justification, is not epistemically self-sufficient.

This present chapter is dedicated to defending this claim with regard to Quine's version of empiricist justificatory monism. First, after introducing the Carnapian position on justification that Quine saw himself as reacting against (Sect. 6.1.1), I detail Quine's version of justificatory monism, and I show that he derives it from verification holism (Sect. 6.1.2). Furthermore, I show that this verification holism presupposes what I call a holophrastic conception of empirical data (Sect. 6.1.3), and I discuss an alternative reading of Quine's verification holism (Sect. 6.1.4). Next, I develop a twofold critique of this line of reasoning (Sects. 6.2.1 and 6.2.2). Finally (Sect. 6.3), I summarize the insights of this chapter by relating the problem diagnosed in Quine's account of justification to Carnap's distinction between external and internal questions.[1]

In Chapter 7, I argue that a similar problem applies to justificatory monism in general, and not only to Quine's variant. Then, I suggest a way how justificatory monists can avoid this problem, one that presupposes scientific realism. Finally, I introduce contemporary scientific realism, and I detail Quine's reasoning for rejecting scientific realism.

In Chapters 8 and 9, I investigate whether Quine was right about this. I examine whether empiricist justificatory monists can defend scientific realism. In Chapter 8, I discuss Stathis Psillos' purportedly scientific arguments for scientific realism, his so-called no-miracle arguments. Then, in Chapter 9, I examine a defense of scientific realism that is not only much more modest and piecemeal, but also much closer to actual scientific research. This defense is due to Penelope Maddy.

[1]As in Chapter 2, my reconstruction of Quine's views is in several ways analogous to both Hylton (2007) and Verhaegh (2015). Where I differ most clearly from both is in my evaluation of these views. In Chapter 3, I have argued that Quine's story of how the infant arrives from stimulus to science is untenable. In the present chapter, I show that the same holds for his attempt to conceive of all Carnapian external questions as basically empirical questions.

6.1 Quine's Case for His Empiricist Justificatory Monism

In this section, I show how Quine's justificatory monism is grounded in his verification holism (Sect. 6.1.2), which is in turn based on what I call his holophrastic empiricism (Sect. 6.1.3). Furthermore, I consider an alternative reading of Quine's verification holism (Sect. 6.1.4). Before doing so, I introduce Carnap's position on how fundamental epistemic norms of science are to be justified. Or better, I introduce Quine's reception of this position (Sect. 6.1.1).

My derivation of Quine's empiricist justificatory monism from his verification holism is similar to Glock (2003, pp. 27–29), Hylton (2007, pp. 76–77), Sinclair (2012), and Verhaegh (2014). My reading differs perhaps most substantially in my emphasis on the pivotal importance of Quine's holophrastic conception of evidence for his verification holism.

6.1.1 Quine's Reading of Carnap's Position on Justification

In his paper *Empiricism, Semantics, and Ontology*, Carnap (1950) develops his famous distinction between internal and external questions, and he introduces the term 'framework'. Carnap (1950, p. 22) sketches his first example of a framework, namely the framework of *The world of things.* Questions internal to this framework, in particular existence questions, permit *bona fide* empirical-scientific answers: they can be "answered by empirical investigations" (Carnap 1950, p. 22). Carnap specifies this as follows:

> Results of observations are evaluated according to certain rules as confirming or disconfirming evidence for possible answers [to internal questions, RG]. (Carnap 1950, p. 22)

Hence, there are rules that determine when a certain observation confirms or falsifies a given answer to an internal question. For instance, there are clear standards to answer the internal question: *Are Unicorns real?* (Carnap's example, ibid.). In contrast, external questions are not scientific questions, even though they may disguise themselves as scientific questions:

> From these questions we must distinguish the external question of the reality of the thing world itself. In contrast to the former questions, this question is raised neither by the man in the street nor by scientists, but only by philosophers. Realists give an affirmative answer, subjective idealists a negative one, and the controversy goes on for centuries without ever being solved. And it cannot be solved because it is framed in a wrong way. To be real in the scientific sense means to be an element of the framework; hence this concept cannot be meaningfully applied to the framework itself. Those who raise the question of the reality of the thing world itself have perhaps in mind not a theoretical question as their formulation seems to suggest, but rather a practical question, a matter of a practical decision concerning the structure of our language. We have to make the choice whether or not to accept and use the forms of expression for the framework in question. (Carnap 1950, pp. 22–23)

In this passage, Carnap maintains that, in contrast to internal questions, external questions concern "the reality of the thing world itself". Carnap is here still referring to his first example of a framework, namely the framework of *The world of things* (Carnap 1950, p. 22). This framework makes possible that internal questions, such as the one asking for the reality of unicorns, are answered in an empirical way. The framework makes this possible by defining clear standards of justification and falsification and by providing a "spatio-temporally ordered system of observable things and events" (ibid.). Asking for the reality of an individual or of a kind of thing in the sense of an internal question is an empirical question precisely because there is this framework that specifies what it means to be real and how to establish the reality of individuals or of kinds in an empirical way.

In contrast, external questions concern the framework itself. An important example for such an external question is whether the framework of the thing world itself is real (Carnap 1950, pp. 21–22). According to the passage, this has been a controversial issue between realists and subjective idealists for centuries—and there is no indication that this debate will ever be resolved. The lack of progress in this debate is no wonder: if the framework itself is in question, then its specifications of what it means to be real, of how to confirm the reality of some entity are unavailable. However, these specifications are presupposed by any factual dispute. Hence, this debate cannot be factual.

Rather than being a factual, or theoretical question,[2] Carnap suggests, philosophers raising the question of the reality of the framework of the thing world are raising a practical question. For Carnap, for a question to be a theoretical question is precisely to be an internal question. Answers to internal questions are subject to definitions and standards such that it is clear what it means to confirm or falsify them. An external question, in contrast, cannot possibly be answered in an empirical-scientific way: it concerns the choice of framework, and only once we have chosen a framework does talk of empirical verification or falsification make sense.

How, then, do you choose an answer to a practical question? In short, Carnap (1950, p. 39) advises to reflect on which framework is most expedient and fruitful for science.

So, the debate between nominalists and realists about numbers should not be decided by deciding the ontological question whether there really are, independently of any framework, numbers. Rather, one should determine which language—the one embodying realist ontological commitments or the nominalist one not including such commitments—contributes most to the flourishing of the mathematical and natural sciences.

Note, finally, that the choice of framework determines which statements are analytic within that framework; or, conversely, that these statements that specify the framework are analytic within that framework. In particular, this includes specifying basic epistemic norms, logical rules of inference, and basic concepts. An example that Carnap (1950) gives is the following: in the framework of the system of numbers, the statement "There are numbers" is analytic because it follows deductively from the statement "Five is a number", which is analytic (Carnap 1950, p. 24). This agrees with Hylton (2007, p. 48), who maintains that the analytic sentences of a language are those that "follow from the rules of the language".

[2]It is potentially confusing that Carnap and Quine use the adjective 'theoretical' in different senses. For Quine, a theoretical statement is ontologically committed and logically structured, and hence contrasted to a holophrastic observation statement. For Carnap, a theoretical statement is one that can be empirically judged, and contrasted to an external, non-scientific statement.

This Carnapian distinction between analytic and synthetic statements is the main target of Quine's programmatic *Two Dogmas of Empiricism*.[3] However, as Richardson (1997) shows, the two philosophers did not understand each other. In particular, Richardson argues, for Carnap, specifying a distinction between analytic and synthetic statements is presupposed by empiricism (or, more Carnapian, by an empiricist language for science)—while Quine thought that a pre-existing conception of empirical evidence, that is, one that does not first have to be specified in terms of analytic statements, undermines the very distinction. Here is how Carnap phrases his conception of empiricism in *Testability and Meaning* (published 1937):

> It seems to me that it is preferable to formulate the principle of empiricism not in the form of an assertion – "all knowledge is empirical" or "all synthetic sentences that we can know are based on (or connected with) experiences" or the like – but rather in the form of a proposal or requirement. As empiricists, we require the language of science to be restricted in a certain way; we require that descriptive predicates and hence synthetic sentences are not to be admitted unless they have some connection with possible observations, a connection which has to be characterized in a suitable way. (Carnap 1937, p. 33)

In this passage, Carnap explicitly suggests to conceive of the principle of empiricism not as an assertion, but rather as a proposal or a requirement for a language of science (the distinction between assertions and proposals in this passage prefigures the distinction between internal and external statements in *Empiricism, Semantics, and Ontology*. Analogously, Carnap's notion of a language for science prefigures the notion of a framework for science).

Furthermore, Carnap cashes out the proposal of empiricism with recourse to the distinction between synthetic and analytic sentences: according to the empiricist proposal for the language of science, synthetic sentences need to have a suitably specified connection to possible observation. In contrast, Quine takes empiricism as given (indeed, after

[3]Roughly, for both Quine and Carnap, synthetic statements extend our knowledge and can be empirically falsified, while analytic statements do not extend our knowledge and are not empirically falsifiable. Compare, in particular, Hylton (2007, pp. 48–50) and Verhaegh (2014, pp. 62–74).

some point in his intellectual development, as *scientifically given*), and then assesses whether there is any place for a distinction between synthetic statements and analytic statements (compare below, Sects. 6.1.2 and 6.1.3).

In sum, the conception of justification that Quine attacks in *Two Dogmas of Empiricism* is that there are two fundamentally different ways of justifying answers to two fundamentally different kinds of questions respectively. External questions, if not mistakenly conceived as internal questions, are to be answered based on practical considerations, such as whether a given answer is "expedient, fruitful, conducive to the aim for which the language is intended" (Carnap 1950, p. 31). These questions concern choices of languages and frameworks. Internal questions, which are theoretical, scientific, or cognitive questions, can only be asked and properly answered once such a framework is in place. In particular, without such a framework in place, it is nonsensical to speak of empirical justification, since the rules specifying empirical justification are furnished by the respective framework. I call this fundamental distinction between the respective ways in which answers to internal and external questions are to be justified *Carnap's justificatory dualism.*

To read Carnap's distinction between internal and external questions as involving two kinds of justification may raise eyebrows of Carnap scholars. It is unclear whether Carnap would call the pragmatic criteria that should guide us in our choice of frameworks as justifying such choices—after all, he does not call answers to such external questions 'assertions'. However, I am here only concerned to delineate Quine's reading of Carnap's position; that this reading is inappropriate in important aspects has been pointed out earlier in this section.

6.1.2 Quine's Verification Holism Underpins His Justificatory Monism

Quine derives his justificatory monism from his verification holism in the following passage from *Two Dogmas*:

> But the total field is so underdetermined by its boundary conditions, experience, that there is much latitude of choice as to what statements to reëvaluate in the light of any single contrary experience. No particular experiences are linked with any particular statements in the field, except indirectly through considerations of equilibrium affecting the field as a whole.
>
> If this view is right, it is misleading to speak of the empirical content of an individual statement – especially if it is a statement at all remote from the experiential periphery of the field. Furthermore it becomes folly to seek a boundary between synthetic statements, which hold contingently on experience, and analytic statements, which hold come what may. Any statement can be held true come what may, if we make drastic enough adjustments elsewhere in the system. Even a statement very close to the periphery can be held true in the face of recalcitrant experience by pleading hallucination or by amending certain statements of the kind called logical laws. Conversely, by the same token, no statement is immune to revision. Revision even of the logical law of the excluded middle has been proposed as a means of simplifying quantum mechanics; [...]. (Quine 1980 [1951], pp. 42–43)

Quine here reasons as follows. He first reaffirms the conclusion of several lines of argument in *Two Dogmas*: the evidence available to us to confirm our overall scientific theory (what Quine figuratively calls 'field' in this passage) underdetermines the theory insofar as a bit of experience that conflicts with our theory does not determine what precise statement of the theory it is contradicting. In *Two Dogmas*, Quine's argument for this famous underdetermination claim is negative. Reductionism (the second dogma of empiricism) has failed, that is, no account has succeeded to show how specific bits of experience are linked to specific bits of theoretical statements.

Hence, it is wrong to suppose that there are one-to-one correlations between specific experiences and specific statements of a theory. Thus, these statements do not have empirical contents on their own, which implies that they are not separately testable. In *Two Dogmas*, Quine boldly claims that it takes "the whole of science" to arrive at a testable empirical prediction (Quine 1980 [1951], p. 42). This means that, as it were, every single bit of evidence is spread across the whole of science, potentially affecting each and every statement of it. I call this position verification holism. Quine never abandoned this verification holism, though he attenuated it somewhat, insofar as he came not to consider the whole of science as the minimal empirically significant unit, but rather *sufficiently inclusive portions* of theory, for example in his article *Five Milestones of Empiricism* (Quine 1981b, p. 71) or in his late *Pursuit of Truth* (Quine 1992, pp. 14–15).

If verification holism is right, then it is misguided to claim that there is a specific class of statements which are true regardless of any empirical evidence. If it is true that every piece of recalcitrant empirical evidence always calls into question entire chunks of theory—including (among others) logical law statements, epistemic norms, and ontological claims—then every statement whatsoever can be called into question by empirical evidence. For instance, if empirical scientists (or philosophers) decide that, given certain evidence, it is fit to abandon the law of the excluded middle, the logician is unable to provide reasons against this that are solely based on logical considerations without reference to experience. Rather, she will have to show that the overall theory is more elegant and still able to accommodate the evidence if the law of the excluded middle is kept in place. Thus, the contribution to successful prediction of empirical data—supplemented with some rough and ready notions of theoretical elegance and simplicity–becomes the fundamental checkpoint for *any claim whatsoever*. This constitutes *Quine's empiricist justificatory monism*.

Hence, Quine draws far-reaching conclusions from his verification holism, that is, the claim that every bit of empirical evidence always corroborates or undermines a sufficiently inclusive portion of scientific theory. Quine urges that verification holism establishes that all statements, including purportedly analytical ones, can be called into question through recalcitrant empirical evidence. There is only a difference in degree, not in kind, regarding the power that empirical evidence has to falsify the law of the excluded middle, as opposed to the power that it has to falsify the clearly empirical hypothesis that there are now 444'444 people living in Zurich.

6.1.3 Quine's Holophrastic Empiricism Justifies His Verification Holism

In this section, I detail the precise conception of empirical evidence that underpins Quine's verification holism and hence also his empiricist justificatory monism. This conception of evidence provides a positive justification for Quine's verification holism, whereas in *Two Dogmas*, Quine only argued negatively.

To begin, let me review how Quine thinks that holistic confirmation of a scientific theory, or, more precisely, of a sufficiently inclusive portion of it, works in detail. To do this, it is necessary to introduce Quine's concept of *observation categorical*. To this point, I have been focusing on experience and observation sentences. As I have shown above (Sect. 2.2.1), observation sentences are distinguished from other occasion sentences by being closely associated with a given pattern of sensory stimulation (as I have also shown, in his later work, Quine did not require anymore that a given observation sentence is associated with roughly the same stimulus meaning *across* different speakers). In *Word and Object*, Quine is content with this vehicle of scientific theory confirmation. It is, however, difficult to see how observation sentences could function as checkpoints for scientific theories. Suppose that X stands for the observation sentence "Red!". Then, a prediction in the form of X does not presuppose any conditions. Thus, a scientific theory that simply predicts X is in no need of experimental verification—indeed in no need of any experiment at all, as its prediction is not tied to any specific precondition. The scientist could simply stretch out by the pool and wait for the verification or falsification of the prediction.

In reality, Quine came to believe, scientific predictions usually come in the form of categorical conditionals. For example: whenever acid is poured on litmus paper, the litmus paper will turn red. Such a structure is what Quine aims at with his observation categoricals. In *Empirical Content*, Quine develops this notion for the first time. He first introduces a variant of his mature conception of observation sentences, requiring only that observation sentences are closely associated to a pattern of sensory stimulations relative to a given subject (see Quine 1981a, p. 25).[4] He then claims that the task of explaining how theory relates to sensory evidence can be recast as the task of "relating theory formulations to observation sentences" (Quine 1981a, p. 25).[5] Quine then argues that this task can be accomplished by joining two observation sentences in a categorical way: whenever observation sentence X obtains, observation sentence Y must obtain as well. Quine calls such a construction 'observation categorical', and Quine (1995, p. 25) conceives them as "a generalized expression of expectation".

To illustrate how Quine thinks that observation categoricals can serve as checkpoints for scientific theory, I now discuss an example that Quine gives for this process. The example occurs in *Pursuit of Truth* (Quine 1992, pp. 9–10) and it is remarkable for its relative complexity. We are invited to imagine a team of mineralogists that encounter an unfamiliar material, which the researchers call 'litholite'. One of the researchers develops a hypothesis of litholite's chemical composition. If the hypothesis is right, the material should emit hydrogen sulfide at a temperature of 180 °C. Due to its very characteristic smell, each of the researchers would immediately recognize hydrogen sulfide. What observation categorical could be used to test the hypothesis? The following should do: "Whenever the material is heated up to 180 degrees Celsius, it emits the characteristic smell of hydrogen sulfide."

How is this observation categorical supposed to link theory to evidence, or the world? Quine (1992, pp. 9–10) gives the following story. There is the body of scientific theory, which Quine (1992, p. 9) imagines as a system of sentences connected by logical laws. In this example, Quine calls this the "backlog of chemical lore". This backlog implies deductively that if the substance has the surmised composition, it will emit hydrogen sulfide when heated over 180 °C. Thus, from the backlog it follows that the observation categorical must hold if the material has the surmised composition. The observation categorical consists of two observation sentences,

[4] He also gives another wording of the goal of naturalized epistemology: its goal is to analyze "the relation between the physicist's sentences on the one hand, treating of gravitation and electrons and the like, and on the other hand the triggering of his sensory receptors" (Quine 1981a, pp. 24–25).

[5] A theory formulation is an axiomatization of a given theory, which contains an infinite number of sentences. For more details, see Quine (2008 [1975], p. 318).

which are, for each of the scientists, directly tied to a specific pattern of sensory stimulation. In our case, the first observation sentence would have a visual pattern as its counterpart, for example some digits on a thermometer (Quine 1992, p. 9). The second observation sentence would correspond to a specific irradiation of the olfactory cells. As a consequence of this direct association of each of the two observation sentences with specific patterns of sensory stimulations, the observation categorical constitutes "an implied generality that the experimenter can directly test" (Quine 1992, p. 10): if litholite is heated up to 180 °C, but the characteristic smell is lacking, the categorical has been refuted, and with it the entire relevant backlog of theory. It is then a matter of pragmatic choice which statements of the backlog to abandon.[6]

Note that, for Quine, a sentence is scientific if it contributes to a theory whose checkpoints are in prediction (Quine 1992, p. 20, this conception of a scientific sentence is analyzed in detail below, Sect. 6.2.2). To contribute here means being an indispensable premise in a valid argument with an observation categorical as its conclusion (compare the example discussed a moment ago). Furthermore, in the more moderate version of Quine's verification holism, such a valid argument constitutes a "sufficiently inclusive portion" (Quine 1981b, p. 71) of scientific theory that has empirical consequences of its own. Hence, this conception of what it means to be a scientific statement is closely connected to Quine's moderate verification holism. This moderate verification holism is in turn based on his holophrastic conception of empirical evidence, as I will now show.

For present purposes, the central feature of Quine's explication of theory confirmation is that these observation sentences, which constitute the observation categorical, relate theories to the world only *holophrastically*:

> More sophisticated observation sentences, including those of specialized scientific communities, are similarly two-faced, even though learned by composition rather than direct conditioning. *What qualifies them as observation sentences is still their holophrastic association with fixed ranges of sensory stimulation, however that association be acquired* [emphasis RG]. (Quine 1992, pp. 7–8)

[6]Quine also declares, following Popper, that such a observation categorical can never be verified *sensu stricto*, but only falsified (see Quine 1992, p. 12).

In this passage, Quine specifies what qualifies a sentence as an observation sentence: in its holophrastic reading, it must be directly associated with a pattern of sensory stimulation.[7] This means that the ultimate checkpoints, the *loci* where theories are falsified, are holophrastic observation sentences that are directly associated with patterns of sensory stimulation. This conception is also explicit in the following passage:

> Man proposes; the world disposes, but only by holophrastic yes-or-no verdicts on the observation sentences that embody man's predictions. (Quine 1992, p. 36)

This means that the ultimate checkpoints for science lack any (proto-) logical structure or ontological commitments. Using Quine's metaphor, one can say that nature does not give its verdict on, and by means of, *bona fide* structured sentences, but only on, and by means of, holophrastic sentences because the contact humans have with the world is restricted to patterns of sensory stimulations. These patterns can be associated directly only with holophrastic observation sentences. I call this conception of empirical evidence *holophrastic evidence*, and sometimes I refer to Quine's empiricism as a *holophrastic empiricism* to emphasize the conception of empirical evidence that is at the core of Quine's empiricism.[8]

It is this holophrastic sort of evidence that is needed for Quine's verification holism to be plausible. To see why, consider the contrast, that is, an epistemology where perception is intrinsically, from the

---

[7] In fact, Quine also requires *intersubjectivity* from any observation sentence, that is, all competent speakers of a community must agree on their verdict on an observation sentence given an appropriate sensory stimulation (see Quine 1992, p. 3). However, this is immaterial for present purposes.

[8] Here is another passage where Quine delineates the same holophrastic conception of scientific evidence. The passage also contains an interesting reflection by Quine on his development regarding the notion of the tribunal of experience: "If there was still an unintended overtone of sensory quality in my reference to surface irritation, it was effectively banished by the time I wrote *Word and Object*; for there I wrote explicitly of the triggering of sensory receptors. [...] Putting matters thus physiologically was of a piece with my naturalism, my rejection of a first philosophy underlying science. Empiricist discipline, however, is not lost thereby. The fabric celebrated in my old metaphor is with us still. As before it is a fabric of sentences accepted in science as true, however provisionally. The ones at the edges are occasion sentences. Moreover, they are occasion sentences of a special sort, namely, ones whose acceptance is prompted by the firing of associated sets or patterns of receptors on that occasion. The tribunal, to worry another of my old metaphors, is just the firing of the receptors" (Quine 1981c, p. 40).

start, (proto-)logically structured and ontologically committed. Such is Tyler Burge's choice (compare above, Sect. 4.2). On this epistemology, for instance, seeing a rabbit would be precisely that: seeing a rabbit, taking it in as a brute given that there is an individual instance of the kind rabbit out there. For Quine, in contrast, seeing a rabbit means being stimulated in a certain way that is associated with an unstructured, holophrastic observation sentence *that we have decided to interpret as* containing a general term, a sortal, that refers to individual rabbits. Adopting the former, Burgean account would mean that non-holophrastic, (proto-)logically structured individual statements with ontological commitments could typically be directly verified or falsified by observation, which in turn implies, even on Quine's own account, that there may be some statements that are confirmed by any experience whatsoever, that is, come what empirical evidence may:

> More directly, the one dogma [reductionism, RG] clearly supports the other [that there is a clear-cut distinction between analytic and synthetic statements, RG] in this way: as long as it is taken to be significant in general to speak of the confirmation and infirmation of a statement, it seems significant to speak also of a limiting kind of statement which is vacuously confirmed, ipso facto, come what may; and such a statement is analytic. (Quine 1980 [1951], p. 41)

According to this passage, distinguishing between synthetic and analytic statements is justified if individual statements are confirmed or rejected by empirical evidence: in this case, it would be possible not only to specify what experiences confirm or undermine a typical empirical sentence, like the empirical claim that there are currently 444'444 people living in Zurich, but also to establish that there are sentences that are confirmed come what empirical evidence may. This would directly contradict Quine's empiricist justificatory holism.

Hence, Quine's verification holism presupposes a very limited conception of empirical evidence, such that the direct empirical test of a (proto-)logically structured, ontologically committed statement is impossible. As I have shown, Quine accounts for this presupposition by maintaining that empirical evidence ultimately consists in patterns of sensory stimulation. These patterns can be directly associated only with holophrastic observation sentences.

Quine's case for his empiricist justificatory monism can be reconstructed as a three-liner:

P1 Empirical evidence is of a kind such that each bit of it always justifies or falsifies a sufficiently inclusive portion of scientific theory.
P2 All scientific statements, including so-called analytic ones, are included in some such portion.
K All statements of science, including so-called analytic ones, can be called into question by an appropriate bit of recalcitrant empirical evidence.

P1 is a specific claim about the nature of empirical evidence, a claim that Quine cashes out with his holophrastic conception of empirical evidence. According to *P1*, empirical evidence always applies to entire portions of scientific theory.

P2 claims that all scientific statements must be part of such a portion that can be empirically tested and thereby potentially falsified. *P2* exploits the equivalence between being a justified statement and being a scientific statement (compare below, Sect. 6.2.2). I have shown earlier in this section that empirical justification or falsification always applies to sufficiently inclusive portions of theory. Hence, a statement that is not part of such a portion cannot be justified. Note that this conception of justified and hence, using the equivalence, of scientific statements merely cashes out Quine's moderate verification holism, which is in turn based on his holophrastic conception of evidence. This means that *P2* can be deduced from *P1*; hence, the only premise that is really needed is *P1*. *P2* is listed to make the reconstruction more explicit.

If *P1* and *P2* are assumed, conclusion *K* follows. *K* asserts that even so-called analytic statements can be called into question by the same kind of evidence by which ordinary empirical statements can be called into question, namely by empirical evidence. Any given so-called analytic statement belongs to an empirically testable portion of theory. Let us assume that it is tested, and the results are not as predicted. Then, the respective bit of recalcitrant empirical evidence is spread over the entire portion of theory to which the respective so-called analytic statement belongs. It is therefore subject to pragmatic considerations which of the statements to abandon in light of recalcitrant empirical evidence. There is no epistemic ground that would prohibit the researchers to abandon the so-called analytic statement in light of this recalcitrant empirical evidence.

This reconstruction can also serve to elucidate the relationship between Quine's empiricist justificatory monism and his checkpoint-empiricism. According to Quine's checkpoint-empiricism, predictions of patterns of sensory stimulations constitute the ultimate checkpoints for science. This phrasing leaves open the possibility that scientific theories contain analytic statements, statements that do not make claims about the world and that are not empirically falsifiable, but that are still indispensable to science. For instance, mathematical statements could be taken to be employed by science, but not falsifiable by empirical-scientific data.

In contrast to this view, Quine's empiricist justificatory monism claims that checkpoint-empiricism reigns supreme: predictions of patterns of sensory stimulations can falsify and justify any statement that is necessary for science, including statements from mathematics or logics.

Note that the conception of the ultimate checkpoints as patterns of sensory stimulation already contains the central presupposition for Quine's empiricist justificatory monism. Hence, thanks to its holophrastic conception of checkpoints, Quine's checkpoint-empiricism implies Quine's empiricist justificatory monism.

6.1.4 An Alternative Reading of Quine's Verification Holism

To conclude my discussion of Quine's empiricist justificatory monism, I discuss a reading of Quine's verification holism, due to Alan Weir, which is diametrically opposed to the one developed here.

To show that Quine's philosophy includes an austere semantic reductionism, Weir argues that Quine subscribes to a verification holism of the following sort:

> Our language divides into theoretical and observational sectors. Sentences in the latter sector have non-null, determinate stimulus meanings, stimulus meanings being scientifically respectable entities characterizable in behavioristic terms. Atypical cases aside, typical theoretical sentences have no empirical consequences on their own (the Duhem-Quine thesis), hence no stimulus meanings. (Weir 2014, p. 129)

Weir suggests here that verification holism divides language into two compartments, one theoretical and one observational. Residents of the latter compartment have a "non-null" determinate stimulus meaning, while members of the former do not have empirical consequences on their own, and therefore no stimulus meanings. This dualistic picture certainly goes against Quine's predilection for gradual distinctions, as Weir (2014, p. 130) admits. However, the crucial claim for this reading of Quine's verification holism lies in the inference contained in the last sentence of the passage above. There, Weir infers from Quine's often made claim that theoretical sentences have no empirical content, or, equivalently, no stimulus meaning *on their own* the much less often made claim that they have *no stimulus meaning at all*.

However, contrary to Weir's reading, Quine holds that theoretical statements, for example those of chemistry, *do* in general have empirical content, as they are part of the "backlog of accepted theory" of observation sentences, see for example, Quine (1992, p. 9).[9] The problem is not that theoretical sentences have no empirical content, but that they have no such content *on their own*—which is the essence of verification holism, as I read it. Such an understanding of verification holism dovetails nicely with Quine's explicit assertions on the issue.

Nevertheless, it could still be that there are systematic reasons that force Quine—*malgré* his own view on the topic—to adopt Weir's notion of verification holism. It seems that Weir holds that there is such a systematic reason, namely Quine's subscription to what he calls "Quine's indeterminacy in the radical form" (Weir 2014, p. 129). According to Weir (2014, p. 129), this indeterminacy thesis implies in particular that in the case of

[9] Analogously, it is completely in line with Quine's overall theory of semantics and with his explicit assertions on the topic that theoretical terms like rabbit do have a stimulus meaning, but one which is not intersubjectively invariant, see, for example Quine (1960, p. 42).

radical translation, it is likely that the sentence-to-sentence translations of two different field linguists will contain mappings that are incompatible with each other. Consider the native sentence "Gavagai". In one manual, it may be translated, for instance, as "There is a rabbit present". In contrast, another field linguist may have decided "Gavagai" means "There is a part of the whole rabbit population existing at that particular moment present". Furthermore, he may have decided that "Gavagai" contains a mass term, not a sortal. Accordingly, he would have to say that "There is a rabbit present" is wrong as a translation of "Gavagai". Hence, Weir can be taken to claim that only his understanding of verification holism can support this indeterminacy thesis.

However, even if it is granted to Weir that Quine subscribes to this strong notion of indeterminacy, my understanding of Quine's verification holism can accommodate this indeterminacy. The field linguist is methodologically speaking in the same situation as the scientist. The evidence that is available to her to set up a translation manual is given by the patterns of sensory stimulations that both the native and the linguist experience. The translation manual that she develops can only be tested by deriving predictions of the following form from it: the native will utter observation sentence *a* when exposed to sensory stimulation of pattern α. In case one of these predictions is falsified, it is unclear which of the sentence-to-sentence translations of the manual are falsified. This is in perfect analogy to the falsification of scientific hypotheses, where it is never clear which part of the 'backlog of theory' has been falsified by a specific experiential result. The reason for this is not that the non-observational sentences of the native receive no empirical content, but rather that it is not determined which empirical content they receive.

To return to the example of the preceding paragraph: "Gavagai", taken as a holophrastic observation sentence, is indifferent to the ontology that is used in its interpretation. For instance, it does not tell the field linguist whether she is confronted with a mass term or with a sortal when confronted with "Gavagai". Thus, I conclude that, in addition to agreeing with Quine's explicit assertions on this issue, verification holism as conceived here can accommodate the evidence that Weir enlists to show the indispensability of his exegetically unappealing understanding of Quine's notion of verification holism.

Let me sum up my reading of Quine's empiricist justificatory monism. In contrast to traditional varieties of empiricism, it rejects the existence of *any* kind of a priori knowledge—including so-called analytic truths, which are warranted a priori, but which do not extend our knowledge of the world in any way. Rather, any statement whatsoever, if warranted at all, is warranted solely empirically. Checkpoint-empiricism reigns supreme. Quine's justification for this empiricist justificatory monism is given by his verification holism. Scientific theories are confirmed, or better, falsified in entire portions, or chunks. The minimal version of such a portion consists in the statements that are required to form the premises of a valid argument whose conclusion is an observation categorical. When deciding whether to modify a chemical, physical, or logical statement of some chunk to accommodate recalcitrant experience, we must rely on pragmatic considerations of elegance and simplicity.

I have shown that this is ultimately because Quine's notion of empirical evidence only allows for the direct falsification of holophrastic observation sentences, or rather, of two holophrastic observation sentences combined to an observation categorical. This makes it impossible to directly falsify (proto-)logically structured, ontologically committed statements and hence leaves the scientist with merely pragmatic guidelines when deciding which theoretical statement of the respective portion of theory has been falsified in the event of a falsification of a (holophrastic) observation categorical. In the next section, I criticize Quine's empiricist justificatory monism.

Before moving on, however, I wish to emphasize the tension between the behavior of the infant that gets conditioned to respond to certain stimuli with certain verbal behavior and the behavior of the scientists in Quine's litholite-example. These scientists are obviously not only responding in a conditioned way to certain stimuli. Rather, they are spontaneously coming up with new hypotheses, which they then test, again in no deterministic, conditioned way. It seems an open question whether there is enough overlap between the infant's verbal behavior and the scientists' spontaneous theorizing and testing to still conceive of both processes as essentially the same one. This, however, is what Quine is committed to, otherwise his entire naturalized epistemology would lose its point, namely, to elucidate the relationship between science and empirical evidence.

6.2 A Critique of Quine's Empiricist Justificatory Monism

In this section, I criticize Quine's justificatory monism in two different ways. First, I briefly summarize my critique of Quine's naturalized epistemology from Chapters 3, 4, and 5, and I show that they question the holophrastic conception of evidence on which Quine's empiricist justificatory monism ultimately stands (Sect. 6.2.1). Second, I argue that Quine cannot conceive of his explication of empirical-scientific justification as itself justified in the sense it itself specifies (Sect. 6.2.2).

6.2.1 Empirical Evidence Is Probably Not Holophrastic

The main target of my critique of Quine's naturalized epistemology in Chapters 3, 4, and 5 is his eschatological physicalism. However, the critique also seriously questions Quine's insistence that the evidence that is available to the infant to learn her first language is holophrastic; that is, that the evidence consists of patterns of sensory stimulations that can be directly associated with holophrastic observation sentences only.

I have first pointed out two problems that are internal to Quine's account. First, I worry that it is not possible to scientifically cash out his invocations of natural selection to explain the intersubjectivity of communication and the success of science. Remember that one of the reasons for Quine to defend the preestablished harmony, accounted for by natural selection, is that he thought it allows him to avoid crediting the infant with perceptions as of objects; he thought that this preestablished harmony allows him to stick with his holophrastic conception of the empirical data available to the infant (see above, Sect. 3.1.4). Second, for his account of how the infant learns her first observation sentences to be minimally plausible, Quine has to credit the infant with understanding of communicative purposes. He effectively does this by claiming that the infant can be pleased by the parent's utterance of approval. However, it is clear that patterns of proximal stimulation, or holophrastic observation sentences, do not furnish information about such communicative purposes.

These two internal problems of Quine's naturalized epistemology provide strong reasons for doubting that Quine has succeeded in giving a scientific account of the genesis of scientific theories that confirms his holophrastic conception of empirical evidence. This is because, in light of these problems, Quine is probably unable to explain scientifically how an infant makes her first steps into language if her only sources of information are given by holophrastic empirical data.

Second, based on research by Tyler Burge, I have developed an external objection to Quine's naturalized epistemology: actual empirical science seriously questions Quine's taking proximal stimulations and holophrastic observation sentences as the basic empirical data that pre-linguistic infants have available. Consider again the case of lightness constancies. As we have seen, bees are able to identify a given distal surface reflectance in spite of drastically different proximal stimulations. Hence, there is solid scientific evidence that such humble beings as bumblebees have more to go on with than proximal sensory stimulations. I have argued (see Sect. 4.3) that Quine should not only accept this finding of perceptual psychology with regard to honeybees, but also with regard to the stimulus that is available to infants to learn their first language, which of course militates against Quine's holophrastic conception of empirical data. In fact, according to contemporary perceptual psychology, the infants' visual system provides them with accurate representations as of environmental particulars with repeatable attributes.

These internal and external objections constitute serious problems for Quine's justification of his holophrastic conception of empirical data. Remember that Quine's naturalized epistemology is Quine's main way to scientifically defend this holophrastic conception of empirical evidence, and hence his specific conception of empirical-scientific justification (which is squarely based on this holophrastic conception of evidence). If successful, his naturalized epistemology would establish that, as a matter of scientific fact, the evidence that humans have to justify their theories is holophrastic. This would in turn justify Quine's verification holism and hence ultimately his empiricist justificatory monism. Hence, my critique of Quine's naturalized epistemology threatens to undermine Quine's verification holism and hence his empiricist justificatory monism.

Note that this line of objection implies that source-empiricism is an empirical doctrine—otherwise it could not be refuted empirically (remember that source-empiricism is the claim that our only source of information about the world is given by patterns of sensory stimulation). This contradicts Keil (2003, pp. 255ff.), who holds that Quine's empiricism *simpliciter* is non-empirical. Since Keil does not distinguish between source- and checkpoint-empiricism, it may be that what he actually wants to claim is that checkpoint-empiricism is non-empirical. With this I agree, as the following section shows.

6.2.2 Quine's Explication of Justification Is Not Justifiable in the Sense It Explicates

In the last section, I have reviewed the results of my critique of Quine's naturalized epistemology and concluded that Quine's conception of the stimulus available to the infant to learn her first language is probably untenable as a scientific hypothesis. In this section, I argue that, for basic structural reasons, Quine's explication of empirical-scientific justification cannot be justified in the way it itself specifies.

In the following passage from *Pursuit of Truth*, Quine reflects on his empiricist conception of science, and on the sort of act that it constitutes to define science:

> But when I cite predictions as the checkpoints of science, I do not see that as normative. I see it as defining a particular language game, in Wittgenstein's phrase: the game of science, in contrast to other good language games such as fiction and poetry. A sentence's claim to scientific status rests on what it contributes to a theory whose checkpoints are in prediction. (Quine 1992, p. 20)

Here, Quine states that the notion of predictions as checkpoints for science is not to be read as a norm, but rather as defining a specific language game in a Wittgensteinian sense of the term: a sentence can claim scientific status insofar as it contributes to predicting future sensory evidence.

What Quine means by 'contributes' here has been detailed above (Sect. 6.1.3): in short, a statement contributes to an empirically testable theory if it is an essential premise in a valid argument, constituted by sentences from the theory, with an observation categorical as its conclusion.[10]

Furthermore, Quine primarily defines what it takes for a sentence to be scientific, namely to contribute to a theory whose checkpoints are in prediction. Only indirectly does he define what makes a theory scientific, namely to have as its ultimate checkpoints prediction. In this passage, Quine uses the rather non-committal term 'prediction' for the checkpoints. However, from the immediate context (see Quine 1992, p. 19), it is clear that he means prediction of sensory stimulation. Hence, the explication under 5 accurately captures this Quinean explication of science:

Explication 5. ***A statement is scientific*** *to the extent to which it contributes to a theory whose checkpoints are in prediction of sensory stimulation.*

When compared with the explication of justification of Quine's empiricist justificatory monism (see above, Sect. 1.1.1), the similarities are striking:

Explication 6. ***A statement is justified*** *to the extent to which it contributes to a theory's ability to predict sensory stimulation.*

Considering explications 5 and 6, it is easy to see that to be justified, for Quine, is largely to be scientific, and vice versa. So, Quine identifies the realm of justified claims (and theories) with the realm of scientific claims (and theories). This expresses the spirit of Quine's justificatory monism, according to which there is only one way of justifying claims, empirical-scientific justification. Note that both explications invoke Quine's holophrastic conception of empirical data: to count as a scientific or as a justified statement, a statement has to contribute to a theory that predicts patterns of sensory stimulation, which are directly associated with holophrastic observation sentences.

Furthermore, consider the relationship between being justified, that is, *contributing to an empirically testable theory*, and *being falsifiable by*

[10] Further textual evidence for reading 'contributes' in this deductive sense is given by passages where Quine uses the term 'imply' to describe the relationship between theory and a certain observation categorical, for example: "The theory will imply a lot of observation conditionals, as I call them, each of which says that if certain observable conditions are met then a certain observable event will occur" (Quine 1981b, p. 70). See also Quine (1992, p. 17).

empirical evidence. I submit that they are two sides of the same coin: any sentence that contributes to such a theory is also a sentence that can be falsified in Quine's holistic sense. First, take a sentence *S* that contributes to a theory that implies a certain observation categorical, and assume that this prediction turns out to be wrong. Then, in Quine's holistic manner, this calls into question the entire relevant part of the theory, including our sentence *S*. Hence, *S* is empirically falsifiable. Conversely, take a sentence that is empirically falsifiable. This must mean that it is part of a portion of a theory that has empirical consequences of its own, since this is the only way how a claim could be empirically falsified. But being a part of such a portion means nothing else than to contribute to it in the way spelled out above: by being an essential premise in a valid argument with an observation categorical as its conclusion.

After the passage quoted, Quine affirms that this language game called science is not committed to "the physical". As a consequence, even telepathy and clairvoyance are conceivable as sources of information for science—in line with earlier assertions, he considers this an issue that can and should be settled by science.[11] However, to establish telepathy and clairvoyance *in fact* as such sources would take an overwhelming amount of evidence, and it would be the end of empiricism. Then, Quine continues to reflect on the consequences of adopting them as such sources:

> Science after such a convulsion would still be science, the same old language game, hinging still on checkpoints in sensory prediction. The collapse of empiricism would admit extra input by telepathy or revelation, but the test of the resulting science would still be predicted sensation.
>
> In that extremity it might indeed be well to modify the game itself, and take on as further checkpoints the predicting of telepathic and divine input as well as of sensory input. It is idle to bulwark definitions against implausible contingencies. (Quine 1992, p. 21)

[11] Note, as elaborated above (Sect. 4.2), that there are much more credible alternatives than clairvoyance and telepathy to Quine's choice of sources of information: if Burge is right, then our visual systems provide us with structured representations as of distal stimuli. This has nothing to do with telepathy, but it flatly contradicts Quine's holophrastic conception of empirical data.

In the first paragraph, Quine states that the language game of science would stay the same, since the checkpoints—prediction of sensory stimulation—would remain the same, even if additional sources of information would be acknowledged.

In the very next sentence, however, Quine reconsiders this assertion: he suggests that in such a case it might even be advisable to change the language game itself by introducing as further checkpoints predictions of telepathic or divine input. Thus, Quine here distinguishes between accepting anti-empiricist elements as sources of information and accepting them as checkpoints. Doing the former amounts to abandoning empiricism, doing the latter amounts to changing the game of science itself. Quine obviously thinks that the collapse of the former does not necessarily imply the collapse of the latter. There is room for manoeuvre here. Hence, in this passage, Quine himself makes the distinction between source- and checkpoint-empiricism that I have repeatedly invoked in this study.

Why should one think that changing the checkpoints of science amounts to a modification of the game of science, rather than to a move in it? In short, because the act of adopting a new standard for a sentence's being scientific will not count as scientific, and hence not as a move in the game, according to the old standard.[12]

To appreciate this in some more detail, consider what evidence it would take to *scientifically* falsify the statement of explication 5, which I call 'HE' in what follows to highlight its intended status as a scientific hypothesis. Now, consider an alternative hypothesis, say *A sentence's claim to scientific status rests on what it contributes to a theory whose checkpoints are in clarity and distinctness* (call this 'HR'). Note that Quine cannot afford to abandon HE without replacing it with another statement specifying the basic rules of the scientific game; otherwise, the game of science would descend into pure anarchy.

So, what could justify abandoning HE in favor of HR? Considering HE, the justification would have to be of this form: *In light of empirical*

[12] It is certainly remarkable that Quine here, at least in passing, remarks that changing the checkpoints amounts to a change of the game of science itself. Does this imply that he himself acknowledges the non-scientific nature of his explication of science? Probably not: note that Quine does not claim that such a change could not be scientifically justified, which is the claim that I am defending here.

evidence E, abandoning HE in favor of HR is going to make our theory more powerful in predicting sensory stimulations than leaving HE in place. But considering the content of HE and HR, it is obvious that replacing HE with HR cannot have a justification of this form.

Note that the justification cannot be that any science that has HR in place will in fact be more efficient at predicting future sensory stimulation than without it—this would presuppose that HE is still in force, as a sort of second-order standard, to which the standard established in HR is subordinate. However, the present concern is with a complete replacement of HE by HR.

This means that a science that is subject to the standard set out in explication 5 cannot falsify the statement of explication 5 itself. To worry Quine's metaphor one last time, the problem is that the officials changing the basic rule of the game of science cannot themselves be bound by this rule—otherwise, they would never change it. Furthermore, it is one of the core features of Quine's empiricist justificatory monism that there must not be a meta-scientific game of epistemological significance. Hence the officials could not be conceived as playing their own game, say, the game of first philosophy, while changing the rules of the object-level game, science. Over and above the game of science there is no other epistemically significant game being played. Everything important takes place within science.

Furthermore, by using the equivalence between a statement's being falsifiable by empirical evidence and its contributing to an empirically testable theory, this means that the statement of explication 5 cannot contribute to any such theory; its claim to the status of science is therefore nonexistent: it is a non-scientific statement. This reasoning can be summarized as follows:

P1 The statement of explication 5 cannot be empirically falsified.
P2 Statements that are not empirically falsifiable do not contribute to an empirically testable theory.
K1 The statement of explication 5 does not contribute to any empirically testable theory.
K2 The statement of explication 5 is unscientific, according to explication 5.

The problem evinced in the case of Quine's explication of science arises, *mutatis mutandis*, with his explication of justification. Remember that being a scientific statement here just means being a justified statement. This implies that the statement of Quine's explication of justification is just as unjustifiable as the statement of his explication of science is unscientific. Just ask yourself what empirical evidence it would take to make it rational to abandon the statement of explication 6, which basically states that the only justification for any claim is given by what it contributes to prediction of sensory stimulations. Again, it is not an option to simply abandon this explication without replacing it: there would then be no explication of justification in force, which would amount to utter and complete epistemic anarchy.

Therefore, let us see what could motivate replacing Quine's explication of justification by a different one. In analogy to the case of science, call Quine's empiricist explication of justification '*JE*', and consider the alternative explication of justification '*JR*': *A sentence is justified to the extent to which it contributes to a theory's clarity and distinctness.* It is hopeless to try to decide between the statements of *JE* and of *JR* by assessing which one of them contributes more to a theory's prediction of sensory stimulation because JR disputes precisely that this is the way how claims are justified. This obviously generalizes to all attempts to replace *JE* by any substantially different explication of empirical-scientific justification. This means that the statement of the explication of justification is not empirically falsifiable.

One can now again exploit the equivalence between *being falsifiable* and *contributing*, and hence *being justified*, and conclude that the statement of Quine's explication of justification is unjustified.[13] This reasoning, which is completely analogous to the one supporting the conclusion that explication 5 is unscientific, can be detailed as follows:

[13]Indeed, it seems that the statement is not only unjustified, but even meaningless. Consider the relationship between meaning and evidence that Quine establishes in *Epistemology Naturalized*: "For epistemology remains centered as always on evidence, and meaning remains centered as always on verification; and evidence is verification" (Quine 1969, p. 89). If meaning is basically verification, an unverifiable statement such as the statement of Quine's explication of justification is meaningless. The same goes for the statements of poetry and fiction: since they do not contribute to a falsifiable portion of theory, they are meaningless.

P1 The statement of explication 6 cannot be empirically falsified in the sense implied by 6.
P2 Statements that are not empirically falsifiable do not contribute to an empirically testable theory.
K1 The statement of explication 6 does not contribute to any empirically testable theory.
K2 The statement of explication 6 is unjustified, according to explication 6.

However, could one not simply distinguish between empiricism as a very general, abstract notion that can be put in many different concrete forms, all of which are then subject to empirical rejection? This is roughly the idea of Bas van Fraassen's proposal of *empiricism as a stance*, and a glance at it will show why this option is not available to the justificatory monist. According to this idea, empiricism as a stance is characterized by a loosely connected group of attitudes which can be put into doctrinal form, but which is not to be identified with one doctrinal form or other, and which is not refuted when some such doctrinal form is refuted (van Fraassen 2002, pp. 47–49). For van Fraassen, these attitudes include a high respect for science as the paradigm of rational inquiry. Furthermore, they require that all factual beliefs are potentially subject to empirical falsification, or as he puts it, that they be delivered "as hostages to fortune, to the fortunes of future empirical evidence, and given up when they fail, without succumbing to despair, cynicism, or debilitating relativism" (van Fraassen 2002, p. 63).[14]

It is clear, however, that the possibility of avoiding the problem at hand by conceiving of empiricism as a stance rather than a theory requires introducing a level of reflection, of belief formation that is distinct from the level of scientific theorizing, while still being highly relevant to it. This extra level saves van Fraassen from the impasse precisely by its being

[14]Thus, the analysis presented here supports van Fraassen's suggestion to conceive of empiricism as a stance rather than a theory because, in the present case, doing the latter amounts to immunizing a specific doctrinal form of empiricism, namely Quine's holophrastic, stimulations-correlated empiricism, against empirical falsification. As a consequence, for van Fraassen, treating empiricism as a scientific doctrine runs counter to the flexible and open-minded spirit of empiricism and has an air of what he dismissively calls metaphysics.

immune in principle against empirical falsification. Of course, such an empirically immune level of justified beliefs is exactly what justificatory monism is fundamentally opposed to.

In short, in this section, I have argued that Quine's explications of scientific sentences and of justified sentences are unscientific and unjustified respectively—in the respective senses of 'scientific' and 'justified' that they themselves specify.[15]

6.3 Quine's Conception of Science Is Not Epistemically Self-Sufficient

In this chapter, I have shown that Quine's empiricist justificatory monism claims that checkpoint-empiricism reigns supreme. That is, Quine's empiricist justificatory monism claims that the only justification for any claim whatsoever lies in its contribution to a theory whose checkpoints are in prediction. This justificatory monism is supported by Quine's verification holism, which is in turn based on a holophrastic conception of empirical data.

I have criticized this line of reasoning in two ways. First, Quine's naturalized epistemology, which is supposed to give a fully-fledged empirical-scientific explanation of how the infant acquires her first language, and thereby to scientifically justify the claim that empirical evidence is indeed holophrastic, probably fails. This removes the main support that Quine provides for this basic presupposition of his empiricist justificatory

[15]These findings contradict a currently common reading of Quine's philosophy. Verhaegh (2015, p. 94) specifies epistemological naturalism in Quine's tradition as rejecting any "detached, science-independent perspective on reality" and instead as aiming at a fully immanent, self-sufficient scientific worldview. However, when discussing Quine's empiricism, Verhaegh (2015, pp. 97–98) fails to make Quine's distinction between source- and checkpoint-empiricism. This leads him to conclude, mistakenly, that what I have called checkpoint-empiricism is subject to empirical refutation, too. In contrast, I have argued here that this can be claimed only for source-empiricism, but not for checkpoint-empiricism.

Note also that it will not help, as Verhaegh (ibid.) does, to argue that this is just a version of Quine's circle at work, that is, of justifying science, as all good naturalists should, by the use of science. I have argued that it is precisely not possible to scientifically justify checkpoint-empiricism; as it were, in the case of checkpoint-empiricism, the circle cannot be closed, whereas it is possible to do this in the case of source-empiricism. However, as I have argued in Chapters 3 and 4, the scientific evidence actually speaks against, and not in favor of Quine's source-empiricism.

monism. Second, I have argued that there is an internal problem in Quine's justificatory monism as such: for basic structural reasons, the statement of the explication of justification at the core of his justificatory monism cannot be justified in the sense specified by this same statement.

This means that empirical science, understood in the Quinean way, is not epistemically self-sufficient: it cannot itself justify its central epistemic norm. This refutes Quine's empiricist justificatory monism; as it were, it has a blind spot at an important point, namely where the basic norm of science is specified.

To conclude this critique of Quine's empiricist justificatory monism, note that this problem of Quine's justificatory monism can be seen as a confirmation of Carnap's position on how basic principles of science are to be justified. In particular, it can be seen as confirming his insistence that specifying a clear-cut notion of empirical justification cannot take place within the practice that the notion itself helps to establish. For the Carnapian, deciding between *JE* and *JR* leads to no fundamental problem: the decision concerns the choice of a suitable framework for science, it is therefore an external matter and has to be decided on a level that is strictly distinct from the level of everyday scientific research. Therefore, Carnap can accept that neither *JE* nor *JR* can be justified in the way which they themselves specify. Somewhat ironically, this means that one way to avoid the predicament of Quine's justificatory monism outlined above (Sect. 6.2.2) is to clearly distinguish between internal and external questions. However, it is precisely this distinction that Quine rejects by rejecting the distinction between analytic and synthetic statements in *Two Dogmas.*

References

Carnap, Rudolf. 1937. Testability and Meaning—Continued. *Philosophy of Science* 4 (1): 1–40.

———. 1950. Empiricism, Semantics, and Ontology. *Revue Internationale de Philosophie* 4 (11): 20–40.

Glock, Hans-Johann. 2003. *Quine and Davidson on Language, Thought and Reality*. Cambridge: Cambridge University Press.
Hylton, Peter. 2007. *Quine*. New York: Routledge.
Keil, Geert. 2003. "Science Itself Teaches". A Fresh Look at Quine's Naturalistic Metaphilosophy. *Grazer Philosophische Studien* 66: 253–280.
Quine, Willard Van Orman. 1960. *Word and Object*. Cambridge: The MIT Press.
———. 1969. Epistemology Naturalized. In *Ontological Relativity and Other Essays*, 69–90. New York: Columbia University Press.
———. 1980 [1951]. Two Dogmas of Empiricism. In *From a Logical Point of View*. Cambridge: Harvard University Press, pp. 20–46.
———. 1981a. Empirical Content. In *Theories and Things*, 24–30. Cambridge: Harvard University Press.
———. 1981b. Five Milestones of Empiricism. In *Theories and Things*, 67–72. Cambridge: Harvard University Press.
———. 1981c. On the Very Idea of a Third Dogma. In *Theories and Things*, 38–42. Cambridge: Harvard University Press.
———. 1992. *Pursuit of Truth*, 2nd ed. Cambridge: Harvard University Press.
———. 1995. *From Stimulus to Science*. Cambridge: Harvard University Press.
———. 2008 [1975]. The Nature of Natural Knowledge. In *Confessions of a Confirmed Extensionalist*, 257–270. Cambridge: Harvard University Press.
Richardson, Alan. 1997. Two Dogmas About Logical Empiricism: Carnap and Quine on Logic, Epistemology, and Empiricism. *Philosophical Topics* 25 (2): 145–168.
Sinclair, Nathan. 2012. A Dogma of Naturalism. *Metaphilosophy* 43 (5): 551–566.
van Fraassen, Bas. 2002. *The Empirical Stance*. New Haven: Yale University Press.
Verhaegh, Sander. 2014. Quine's Argument from Despair. *British Journal for the History of Philosophy* 22 (1): 150–173.
———. 2015. Rafts, Boats, and Cruise Ships. Naturalism and Holism in Quine's Philosophy. PhD thesis, University of Groningen.
Weir, Alan. 2014. Quine's Naturalism. In *A Companion to W.V.O. Quine*, ed. Gilbert Harman and Ernie Lepore, 114–147. Chichester: Wiley Blackwell.

7

Why Justificatory Monism Needs Scientific Realism

In this chapter, I suggest that a problem similar to the one diagnosed for Quine's empiricist justificatory monism generalizes to all justificatory monisms (Sect. 7.1). Then, I introduce a kind of justificatory monism, what I call realist justificatory monism, which presupposes scientific realism. Furthermore, I argue that this realist justificatory monism can come to terms with said problem (Sect. 7.2). I conclude by introducing contemporary scientific realism together with Quine's rejection of this doctrine (Sect. 7.3).

7.1 The Problem of Justificatory Monism: Justificatory Pluralism

In the preceding chapter, I have reconstructed and critically assessed Quine's empiricist justificatory monism. This doctrine asserts that the only way of justifying any claim whatsoever is given by its contribution to prediction of holophrastic empirical evidence. I have then developed two objections against this doctrine. The second objection argues that the statements specifying two central (and largely identical) conceptions of

R. Gubelmann, *A Science-Based Critique of Epistemological Naturalism in Quine's Tradition*, https://doi.org/10.1007/978-3-030-24524-5_7

Quine's own epistemology, namely *scientific statement* and *justified statement*, are unscientific and unjustified in the respective senses that they themselves specify.

A similar problem applies to justificatory monism in general: the basic problem of any given instance of justificatory monism is that it cannot reject any alternative conceptions of justification without begging the question. For any two conceptions of justification, *J1* and *J2*, proponents of *J1* will always observe *J1* when defending *J1* against *J2*, while proponents of *J2* will observe *J2* to defend *J2* against *J1*. Both of these defenses are begging the question, since the target audience of the respective defenses, proponents of *J2* and *J1*, respectively, do exactly not accept justifications along the lines specified in *J1* and *J2* respectively.

For instance, consider the case where *PE*, a proponent of *JE*, according to which a sentence is justified to the extent to which it contributes to a theory's ability to predict sensory stimulation, gets into an epistemological debate with *PR*, a proponent of *JR*. According to *JR*, a sentence is justified to the extent to which it contributes to a theory's clarity and distinctness. *PE* is committed to defend *JE* by pointing out that *JE* contributes massively to a theory whose checkpoints are in prediction of sensory stimulation.[1] Unfortunately, such a defense will not impress *PR*, since she thinks that a claim is justified by its contribution to clarity and distinctness. The converse holds as well: *PR*'s detailed case to show how wonderfully *JR* contributes to a theory's clarity and distinctness must leave *PE* entirely unimpressed.

Note, furthermore, that it will not help if proponents of each of these justificatory monisms develop their corresponding naturalized epistemologies. These would be scientific accounts, in the sense of scientific specified by their respective conceptions of empirical-scientific justification, of the genesis of scientific theorizing; by the same token, they would confirm their respective conceptions of how empirical-scientific justification works. There would then be as many naturalized epistemologies as justificatory monisms, and the problem would once more be that each of

[1] I have argued above, Sect. 6.2.2, that this is not possible. The debate envisaged here is solely for illustrative purposes.

these naturalized epistemologies would only be accepted by adherents of the corresponding explication of empirical-scientific justification.

For instance, assume that Tyler Burge's work in *Origins of Objectivity* (Burge 2010) is proposed as such a naturalized epistemology to support Burge's explication of empirical-scientific justification (whatever that explication may be). To the Quinean justificatory monist, however, this work would in many aspects not qualify as scientific, in particular because it irreducibly posits intentional entities. Hence, Quine could not be converted to Burge's explication of empirical-scientific justification by this Burgean naturalized epistemology.

This phenomenon is a consequence of the relationship between naturalized epistemology and given versions of justificatory monism (see above, Sect. 1.1.3): any given version of justificatory monism determines which modes of inquiry are admissible for naturalized epistemology (which should then in turn empirically corroborate the respective version of justificatory monism).

This problem would evaporate if both parties could negotiate their disagreement on a different level of discourse, using a different explication of justification to which both parties agree. However, being justificatory monists, neither of the two parties will accept such a level of discourse with such a different explication of justification. Rather, both of them insist that there is only one way of justifying claims—which both parties understand differently. As a consequence, neither of the two parties has any chance to convince the other.

It is misguided to assume as justified a hypothesis for which there exist many alternatives, none of which one can eliminate. This is exactly the case here: A proponent of justificatory monism based on *J1* is unable to say, without immediately begging the question, why *J1* is the correct explication of empirical-scientific justification, but not *J2*, or *J3*, etc. Therefore, justificatory monists cannot justify the respective explication of justification with which they operate. I call this the problem of *justificatory pluralism*. The structure of this reasoning can be reconstrued as follows:

P1 There is a great number of different explications of scientific justification.

P2 Each of these explications yields a different version of justificatory monism.

P3 Proponents of each of these versions of justificatory monism are committed to observe their respective explication of justification when arguing why their specific version of justification is superior to competing explications of justification.

K1 All such arguments are begging the question: they presuppose exactly what is not accepted by their dialectical counterpart, namely the superiority of their respective explication of justification.

K2 No justificatory monist can show that her specific explication of justification is superior to any of the numerous competing explications.

K3 None of the numerous explications of justification can count as justified.

Note that Maddy cannot avoid this problem by emphasizing that her justificatory monism is implicit. While this aspect of her position might save her from entering into question-begging debates about explicitly held conceptions of empirical justification, she still faces the essence of the problem, no matter how implicit her commitment to empiricist justificatory monism. If it is the case that there are numerous substantially different conceptions of empirical-scientific justification, each being epistemically isolated from all others, then the Second Philosopher would be implicitly committed to one of these conceptions, being unable to say why she employs this one rather than any of the other conceptions. Furthermore, she would be forever epistemically bound to this one conception of justification, as she could never justify changing her implicit allegiance from one of these conceptions of justification to the other. Hence, while she would not end up in a fatal dialectical deadlock with proponents of other conceptions of justification, she would be stuck, equally fatally, with one such conception, and her implicit allegiance to it would be entirely arbitrary.

7.2 How a Realist Justificatory Monism Can Avoid the Problem

What is needed for justificatory monists to escape from this dead-end is a court of appeal, a tribunal, that is (1) acceptable to justificatory monists and (2) able to decide, at least in principle, which of the various conceptions of justification should be adopted. Such a tribunal could break the dialectical deadlock between proponents of incompatible explications of justification, a deadlock that implies that none of the various conceptions of justification is justified.

A candidate for this tribunal would of course be first philosophy, as Quine conceives it: a "supra-scientific tribunal" that could decide between competing conceptions of scientific justification (Quine 1981a, p. 72, compare also above, Sect. 1.1.1). Of course, it is clear that such a supra-scientific tribunal fundamentally contradicts justificatory monism as such.

There is, however, a tribunal that justificatory monists may find acceptable: the tribunal of nature itself. We have seen above (Sect. 6.1.3) that Quine thought that the world delivers its verdicts only in holophrastic yes-or-no sentences. But what if Quine was wrong? What if the world, or nature, determines the right explication of justification? There would then be an objective fact of nature, independent of our conceptualizations of it, that determines which explication of justification is the right one. In other words, the correct explication of justification would be given by reality, to be discovered and accurately represented by science just like any other part of reality. I call this *realist justificatory monism.*

Realist justificatory monism requires that the goal of science is no more understood in Quine's empiricist way, but rather in a thoroughly realist way: the goal of science is to accurately represent the structure of the world, as it exists independently of any theories or conceptualizations. The correct explication of justification is then just one part of the world that science intends to represent accurately. Hence, realist justificatory monism also requires that reality has a mind- and theory-independent structure, and that science is typically able to accurately represent this structure. These two last requirements jointly constitute contemporary scientific realism (see below, Sect. 7.3.1).

Given scientific realism, the problem of alternative and incompatible conceptions of justification, such as *JE* and *JR*, would no more threaten the very essence of justificatory monism. Even if, at present, we are unable to discern whether *JE* or *JR*—or none of them—is the correct explication of justification, we know that such a correct conception exists, namely the one given by nature itself. Furthermore, scientific realism gives us reason to be optimistic that we will sooner or later be able to identify this correct conception.[2] Interestingly, on this conception, the tribunal of nature would then gradually reveal *itself* to the scientists. Putting the metaphor aside, this can be spelled out as follows. The correct explication of justification, the one that can be found in nature itself, would also be the epistemic means by which the scientists could ascertain that they have indeed found the correct explication of empirical-scientific justification. For instance, assume that nature's own explication of justification is *JN*: *A hypothesis is justified if it successfully makes novel predictions*. Then, the only way how scientists can be sure that *JN* is the correct explication of justification is by verifying that it actually fulfills *JN*, that is, that *JN* itself successfully makes novel predictions. This is not viciously circular if we presuppose scientific realism; or rather, given scientific realism, the circle would be so broad as to include nature itself.

It is true that scientific realism has traditionally been claimed mostly with regard to kinds of material entities, such as atoms, but there seems to be no reason in principle why it should not be applied to non-material structures as well. In particular, Burge (see above, Sect. 4.2) makes a convincing science-based case for the existence of natural perceptual norms.

Furthermore, scientific realism can plausibly provide a solution to the Second Philosopher's variety of the problem of justificatory pluralism. If the concepts of scientific theories typically match natural kinds with increasing precision as science progresses, the Second Philosopher could assume that the same process would cause the numerous conceptions

[2] This scenario requires that proponents of specific conceptions of justification, say *JE*, need to be able to recognize instances of the correct explication of justification as such instances even if they may contradict their respective conceptions, say *JE*. To explain how this is semantically and psychologically possible, one could adopt a semantics along the lines specified by Burge (2012; 2013, pp. 517–520). In these articles, Burge sketches a semantics where speakers of a language are able to apply a term, such as 'justified', without having a clear-cut explication of justification, or they may even apply it in a way that contradicts their explicitly held conceptions. They are able to do so because they have other means than fully-fledged conceptions, what Burge (2013, p. 518) calls "patterns of first-order usage", to decide whether it is appropriate to use a certain term in a certain situation.

of empirical-scientific justification, including the one that she happens to be implicitly and, for the time being, arbitrarily, committed to, to converge. This convergence would rationalize the Second Philosopher's implicit commitment and disclose a real possibility to resolve justificatory pluralism without having to resort to meta-scientific modes of discourse, modes that the Second Philosopher wants to avoid at all costs.

In summary, given scientific realism, nature can constitute a tribunal with the authority and the communicative channels to decide which of the competing conceptions of justification should be adopted. Thus, it can break the fatal dialectical deadlock of competing and hence unjustified conceptions of empirical justification. Hence, the possibility of an empiricist justificatory monism depends on whether empiricist justificatory monists can justify scientific realism.

7.3 Scientific Realism: A Quinean Perspective

In this section, I urge that, *pace* a number of contemporary Quine scholars, Quine never embraced scientific realism, and I argue that he is right not to have done so. First, I introduce contemporary scientific realism following the lead of one of its most prominent defenders, Stathis Psillos (Sect. 7.3.1). Then, I show, in Sect. 7.3.2, that the doctrine is fundamentally at odds with Quine's inscrutability of reference, a position that Quine maintained until the very end of his publishing career. Finally, I suggest that Quine's holophrastic conception of evidence which, as I have shown above (Sect. 6.1.3) underpins his empiricist justificatory monism, also implies the inscrutability of reference. Hence, scientific realism threatens the very basis of Quine's empiricist justificatory monism.

7.3.1 Contemporary Scientific Realism

In this section, I detail the contemporary conception of scientific realism. According to Psillos, the essence of scientific realism is given by the following three claims (Psillos 2009, p. 34):

The Metaphysical Thesis:	The world has a definite and mind-independent structure.
The Semantic Thesis:	Scientific theories should be taken at face value.
The Epistemic Thesis:	Mature and predictively successful scientific theories are (approximately) true of the world. The entities postulated by them, or, at any rate, entities very similar to those posited, inhabit the world.

Of these three claims, the third one is most important. Psillos calls it an epistemic one. In addition to stating a claim regarding the truth of scientific theories, which is discussed in detail below, it does make a stronger metaphysical claim than the first: it claims that the world is inhabited at least by those entities postulated by mature and predictively successful scientific theories. The first of the three theses makes a metaphysical claim about the structure of the world, namely that it has a determinate and mind-independent structure. What does it mean that the world's structure is mind-independent? Psillos (2009, p. 40) understands mind-independence as opposed to the position of modern verificationists, according to which what exists in the world is determined by what we can know about the world. Against this position, mind-independence maintains that the world ultimately passes its verdict on our conceptualization of it: "At stake is a robust sense of objectivity, namely, a conception of the world as the arbiter of our changing and evolving conceptualizations of it." To emphasize the world's independence of our conceptualizations of it, I sometimes also use the term 'theory-independent': what exists in the world is not determined by our theories about the world. This first claim clarifies the third one as follows: to say that some entities postulated by scientific theories "inhabit the world" means that the respective theories correctly represent the respective part of the mind-independent structure of the world.

The second urges that scientific theories should be taken at face value. What is the face value of a scientific theory? Considering other assertions of Psillos, it turns out that he uses this second thesis to spell out what he means by his claim, made in the epistemic thesis, that scientific theories are approximately true of the world. Psillos (2009, p. 4) contrasts it to eliminative instrumentalist and reductive empiricist accounts, where the main contrast is that, on these accounts, the truth of a given scientific statement does not depend on whether the entities postulated by it do exist mind-independently. Rather, the statement is reduced to some ontologically lightweight observational claim; if this latter claim is corroborated by experiential methods, the original statement is counted as true. In contrast, those who take scientific theories "at face value" will, in the first instance at least (see Psillos 2009, p. 42), count such a statement as true only if the posited entities exist in the world. Thus, this second thesis is intended to make plain that semantic considerations, indeed, a correspondence theory of truth, are inextricably intertwined with realism; as Psillos (2009, p. 42) puts it, "Semantics (and truth) enters the realist position from the front door, by issuing a literal understanding of the existential assertion".

In the following, I argue that this is wrong: semantic as well as epistemic considerations are connected in important ways with scientific realism, but they are not at the core of the doctrine.

Psillos takes pains to establish that scientific realism is not *only* a metaphysical doctrine. He writes that scientific realism does not only concern the world, but also scientific theories. On the following page, Psillos elaborates on this twofold nature of scientific realism by linking it to what he calls a "factualist conception of reality". Implicitly, it seems that this conception amounts to conceiving of reality as the totality of facts. Together with a correspondence theory of truth, this factualist conception yields what Psillos calls a "two-way traffic between truth and reality":

> Reality is the realm of facts (truth-makers) and to say that a representation of it is true is to say that it represents a fact: we can go from truth to the facts and from the facts to truth. (Psillos 2009, p. 35)

According to this passage, reality is the totality of truth-making facts and, conversely, a claim is true if it represents a fact. Furthermore, there is a qualitative difference between truth and reality: "[t]ruth is attributed to our *representations* of the world. Reality is attributed to the world." (ibid.). Accordingly, an important difference between facts and claims is that the former can be real, while the latter can be true, but not vice versa. When elaborating on his notion of correspondence, Psillos writes that according to his view, a statement is true "if the referred-to entities stand in the referred-to relations" (Psillos 1999, p. 244).

Before arguing that accepting such a correspondence theory of truth is neither a sufficient, nor a necessary condition for the acceptance of scientific realism, note that the correspondence relation itself must be admitted among possible "referred-to relations", since Psillos certainly wants to claim that the statement "theory T is true" applies to some theories (indeed he wants it to apply to all mature and predictively successful theories, compare below, Sect. 8.2.2). But if this statement is true, then there must be a fact to which it corresponds, and that fact must be real. The fact in question would be that the theory T stands in the correspondence relation to another fact. This contradicts the clear-cut distinction between facts, which are real, and our representations, which are true. The theory T is at the same time part of a fact and something that is true.[3]

Showing that adopting a correspondence theory is not sufficient to accept scientific realism is rather simple. As Rasmussen (2014, pp. 12–17) argues convincingly in his detailed defense of correspondence theories of truth, one can easily imagine a correspondence theory that has nothing whatsoever to do with mind-independent facts. For instance, one could propose that a statement is true if it corresponds to certain phenomenologically conceived sense impressions.

A correspondence theory of truth only points into the scientific realist's direction if it contains the metaphysical claim that is at the center of scientific realism: adopting a correspondence theory of truth is sufficient to adopt scientific realism only if the relevant correspondence relation is specified as a correspondence between claims and mind- and theory-independent facts. Conversely, rejecting scientific realism commits one to reject a *specific* correspondence theory of truth, namely one that requires of a claim correspondence to the mind-independent structure of the world to be true.

Furthermore, it is not a necessary condition either. As Psillos himself acknowledges, several researchers have claimed that scientific realism is best conceived as a primarily metaphysical doctrine. According to Devitt

[3]Thus, applying the correspondence theory to scientific realism furnishes a first hint that scientific realism somehow differs from traditional scientific theories: it leads to an anomaly in a factualist correspondence theory.

(1997, p. 47), strong scientific realism, which is the kind of scientific realism that Psillos wishes to defend, boils down to the following claim: "Tokens of most unobservable scientific types objectively exist independently of the mental, and (approximately) obey the laws of science". With regard to this purely metaphysical understanding of scientific realism, Psillos (2009, p. 41) worries that "existence", as used in Devitt's characterization, is ambiguous, and that a correspondence theory is required to resolve the ambiguity in the way preferred by the realist. Here is how Psillos phrases his concern:

> As noted above, a modern verificationist can (and does) accept that electrons exist. Their gloss on *existence* is that it does not make sense to talk about the existence (or reality) of electrons unless we understand this assertion to mean that ..., where the dots are replaced with a suitable epistemic/conceptual condition. (Psillos 2009, p. 41)

To avoid such anti-realist readings of, say, Devitt's definition of strong scientific realism, Psillos argues, it must be added that the conditions for attribution of existence must not be epistemic or conceptual. Furthermore, Psillos argues, it is not semantically innocent simply to claim that all the scientific realist has to do is to take the respective claims *at face value.* Instead, taking these claims "at face value" requires according to Psillos, as shown above, that one adopts a specific theory of truth. Even though Psillos is not explicit about it, this is intended to show that a correspondence theory of truth must be adopted by anybody who adopts scientific realism.

Unfortunately, this reply is not convincing. Hilary Putnam, one of the most important critics of scientific realism as understood here, will not reply to the claim of the strong realist, according to which, say, electrons exist independently of our intellect: "I wholeheartedly agree, just don't think that I am committing myself to any ontological claims by doing so." Rather, he would say that he disagrees *because we cannot know whether these things exist in this sense,* and that all we know is that they have passed all of our empirical tests with admirable results. Putnam aptly phrases his main reservations against scientific realism by characterizing it as seeking "a description which 'corresponds' to the noumenal facts" (Putnam 1990, p. 40). Like Kant, Putnam holds that we have no epistemic ground to stand on when deciding whether we have such a description. What we do have here are *epistemological* reasons against a certain *metaphysical* position, namely that the posits of mature and successful scientific theories exist in mind-independent nature. This, however, does not mean that scientific realism itself is an epistemological (let alone a semantic) position; rather, it means that one can come to accept or reject the position on semantic and epistemological grounds.

Psillos effectively confuses two issues in his relentless defense of a correspondence theory of truth. He is right when he urges that realists must distinguish between a scientific community's acceptance of a given theory and that theory being an accurate representation of the structure and furniture of nature (see Psillos 2009, p. 36). He is wrong, however, in his insistence that, to do so, realists must reserve the truth predicate for theories that provide such an accurate representation. As such, it does not contradict realism to apply the truth-predicate to theories that are simply accepted by the relevant community. Such a so-called epistemic conception of truth contradicts realism only if it comes with an *epistemological* argument to the conclusion that there are no theories which, in addition to being true in this epistemic sense, also accurately represent the furniture of nature.

Egg (2014, pp. 8–9) attempts to combine Psillos' insistence that metaphysics is not clearly dominant in the position of scientific realism with Devitt's insistence that it is. On the one hand, he suggests that Psillos would not deny that scientific realism is "first and foremost a metaphysical". On the other hand, Egg suggests that even Devitt would admit that scientific realism can be challenged in a way that requires a semantic or an epistemic response. The textual evidence cited here testifies for Psillos' willingness to deny the first claim; however, the analyses made here support Egg insofar as Psillos' variant of scientific realism is *in fact* first and foremost metaphysical. Furthermore, the second claim is warranted as long as it is kept in mind that the content of a claim is one thing, while the way of defending it is another matter.

Finally, note that scientific realism is usually contrasted with realism *simpliciter* with its limited scope: scientific realism usually feels no need to defend the existence in reality of everyday, ordinary, visible objects, but rather focuses on so-called theoretical, unobservable entities like electrons, neutrinos, etc. There is no canonical way to draw the distinction, no agreed-on set of theoretical entities that are at issue for the scientific realist, but it is helpful nonetheless to keep this in mind. A contemporary scientific realist feels no need to respond to skepticism regarding the external world, and similar issues (usually because she thinks that these skepticisms are bankrupt); she merely engages, as it were, skepticism regarding unobservable, theoretical entities.

To sum up, I have introduced scientific realism, urging that it is primarily a metaphysical doctrine. It makes a claim about mind- and theory-independent reality, namely that this reality is populated by entities that correspond to the ontological posits of (a subset of) contemporary scientific theories. This is not to say that this doctrine is independent of epistemological considerations. In particular, scientific realism is committed to the epistemological claim that human beings are capable of developing scientific theories that accurately represent reality.

Furthermore, adopting a correspondence theory of truth is neither necessary nor sufficient for accepting scientific realism. However, proposing a correspondence theory of truth with a specific, realistically-tilted correspondence relation, together with the claim that some scientific theories are true in this sense, implies scientific realism. In proposing such a correspondence theory, Psillos is able to phrase the doctrine of scientific realism in an elegant way: scientific realism claims that some scientific theories are true. As such, there is no problem with putting scientific realism in this way. In the discussion of Psillos' and Sankey's arguments for scientific realism (see below, Chapter 8), their use of such a realistically-tilted correspondence theory will be adopted. However, to emphasize the metaphysical substance that these authors put into their conception of truth, I sometimes use the term 'metaphysical truth' or 'correspondence truth'.

7.3.2 Quine's Case Against Scientific Realism

In this section, I argue that Quine did not embrace scientific realism, as detailed in the preceding section. The strongest evidence to the contrary stems from the following intriguing passage from "Natural Kinds"[4]:

> A sense of comparative similarity, I remarked earlier, is one of man's animal endowments. Insofar as it fits in with regularities of nature, so as to afford us reasonable success in our primitive inductions and expectations, it is presumably an evolutionary product of natural selection. Secondly, as remarked, one's sense of similarity or one's system of kinds develops and changes and even turns multiple as one matures, making perhaps for increasingly dependable prediction. And at length standards of similarity set in which are geared to theoretical science. This development is a development away from the immediate, subjective, animal sense of similarity to the remoter objectivity of a similarity determined by scientific hypotheses and posits and constructs. Things are similar in the later or theoretical sense to the degree that they are interchangeable parts of the cosmic machine *revealed* by science [emphasis RG]. (Quine 1969, p. 133)

In the first part of the passage, Quine reminds readers that infants have an innate sense of perceptual similarity, and that this sense guides us to successful primitive inductions; as he puts it earlier, the sense is helpful to predict important events on the "food-gathering level" (Quine 1969, p. 127). Quine contrasts this innate sense of similarity with the similarity standards of theoretical science. These latter standards are the end-point of a development away from subjective, animal similarity towards objective similarity. Quine claims that to be similar in this latter, objective sense means to constitute interchangeable parts in the cosmic machine. Finally, what is central for present purposes: Quine writes that this machine is not posited or invented, but revealed by science.

---

[4]Thus, Kornblith (1993, pp. 5–7) emphasizes this article for his reading of Quine as a scientific realist. This chapter contests Kornblith's claim that a thoroughly realist, essentialist notion of natural kinds can play a key role in "filling out the epistemological project which Quine himself developed" (Kornblith 1993, p. 7), since, as I will urge, it is essential to Quine's naturalism not to be combined with anything like contemporary scientific realism. Castro (2013, p. 45), whose views on the atoms debate are critically assessed in Sect. 9.2.3 below, also reads Quine as a contemporary scientific realist.

Obviously, something that is revealed was there beforehand. Hence, if science reveals the cosmic machine, it describes a structure that exists before, and outside the scientific descriptions themselves. That science is typically, or at least in certain cases, able to do this is the core claim of contemporary scientific realism, as I have introduced it in the preceding section.

In the remainder of this section, I defend the claim that this passage should not be taken to imply that Quine accepts scientific realism. I do so by detailing Quine's doctrine of the inscrutability of reference, which, as I will urge, directly contradicts scientific realism. Furthermore, I argue that Quine's favored way to justify the inscrutability of reference from the sixties onward was by means of his holophrastic conception of empirical data. This conception is also the ultimate basis for Quine's justificatory monism (see above, Sect. 6.1.3). Hence, rejecting the inscrutability of reference implies, by *modus tollens*, rejecting this holophrastic conception of empirical data. This removes the main justification that Quine gave for his justificatory monism. This outweighs the evidence given for scientific realism by the passage from *Natural Kinds*, which, I will urge, should be read in the sense of Quine's immanent realism.

In the following, I close-read central passages from Quine's *Things and Their Place in Theories.* There are other passages that support the gist of what he is relating in this passage (I will refer to some of them in passing through the passage), but it is the only passage in his writings published after *Word and Object* that details the justification for his inscrutability of reference and relate the doctrine to his so-called robust realism on the one hand and to his rejection of first philosophy on the other. For simplicity reasons, I have numbered the quoted paragraphs.[5]

> [1] It is occasion sentences that report the observations on which science rests. The scientific output is likewise sentential: true sentences, we hope, truths about nature. The objects, or values of variables, serve merely as indices along the way, and we may permute or supplant them as we please

[5] Here is a passage from Quine's late *Pursuit of Truth* that parallels the passage of paragraph (1): "[w]hat particular objects there may be is indifferent to the truth of observation sentences, indifferent to the support they lend to theoretical sentences, indifferent to the success of the theory in its predictions" (Quine 1992, p. 31).

> as long as the sentence-to-sentence structure is preserved. The scientific system, ontology and all, is a conceptual bridge of our own making, linking sensory stimulation to sensory stimulation. I am repeating what I said at the beginning. (Quine 1981b, p. 20)

In this first paragraph, Quine emphasizes the pivotal importance of sensory stimulation and of occasion sentences that, in their holophrastic use, are directly associated with patterns of sensory stimulation for science. Quine also highlights the relative insignificance of the ontological posits of science for scientific inquiry. Science is based on observation reports in the form of occasion sentences (observation sentences are a kind of occasion sentences; occasion sentences are true on one occasion, but false on others, see above, Sect. 2.2.1). Furthermore, the products of scientific theorizing are again sentences. Hence, there are no epistemic restrictions against replacing or permuting the ontological posits of these sentences as long as the occasion sentences in their holophrastic use are not affected.

Quine's most famous examples of replacing a theory's ontology with a different one that does not violate any evidence are the so-called proxy functions. Proxy functions are mappings of a given ontology of a theory onto a very different one such that the resulting theory remains unchanged with regard to its holophrastic observation sentences (that provide the empirical support for the theory) and with regard to the theory's basic structure, compare Quine (1981b, p. 19). One example for such a proxy function is a mapping of our physical objects on the placetime that they each occupy, that is, on spaciotemporal quadruples. This proxy function replaces physical entities with abstract structures. According to Quine, there is no empirical basis to prohibit this mapping.

Quine (1981b, p. 19) calls the doctrine, according to which all the evidence that we could possibly muster provides no reason against replacing our overall ontology with a completely different one via such a proxy function *inscrutability of reference.* This inscrutability of reference certainly contradicts the basic thrust of scientific realism, according to which there is only one way the world is structured, and successful scientific theories typically represent this structure accurately. How then can Quine maintain any kind of commitment to what he calls a robust realism, given the inscrutability of reference? Quine clarifies this as the passage continues:

> [2] But I also expressed, at the beginning, my unswerving belief in external things – people, nerve endings, sticks, stones. This I reaffirm. I believe also, if less firmly, in atoms and electrons and in classes. Now how is all this robust realism to be reconciled with the barren scene that I have just been depicting? The answer is naturalism: the recognition that it is within science itself, and not in some prior philosophy, that reality is to be identified and described.
>
> [3] The semantical considerations that seemed to undermine all this were concerned not with assessing reality but with analyzing method and evidence. They belong not to ontology but to the methodology of ontology, and thus to epistemology. *Those considerations showed that I could indeed turn my back on my external things and classes and ride the proxy functions to something strange and different without doing violence to any evidence* [Emphasis RG]. But all ascription of reality must come rather from within one's theory of the world; it is incoherent otherwise. (Quine 1981b, p. 21)

In this important passage, Quine first specifies, in paragraph (2), what he means by 'robust realism'. Quine writes that he unswervingly believes in external things, in spite of his insistence that all these external things are theoretical posits that we could replace by "something strange and different" through a proxy function. It is essential to Quine's robust realism that it incorporates this insight. Such a position is inconsistent with contemporary scientific realism: scientific realists are precisely contesting that kinds such as atoms or electrons are real only in the sense that they are posited by a predictively highly successful theory: they insist that they are part of the furniture of the world *in addition to being part of our theories.* Quine's robust realism is restricted to the first assertion.[6]

Furthermore, Quine emphasizes that this combination of belief in external things with belief that our evidence does not determine that there are material objects rather than quadruples of numbers is underwritten by his naturalism. This naturalism is described as based on the insight that reality must be described from within science, and not in some 'prior philosophy', a term that Quine uses interchangeably with 'first philosophy' (remember that I call 'justificatory monism' what Quine calls his 'naturalism').

[6]This reading of Quine's realism is congenial to Ben-Menahem (2016), who urges that Quine, like James, rejected contemporary scientific realism.

Hence, for Quine, it is a consequence of the rejection of first philosophy that realism cannot amount to more than the robust or immanent realism that he has just introduced: unless we have another kind of access to reality than the one provided by empirical science, we cannot discern whether the posits of our most successful theories are real in a sense other than being such posits. Epistemically speaking, there is no other way to justify a posit than the empirical-scientific way. If this way shows that other posits would be just as justified, as Quine thinks is the case, we have to acknowledge this, but we are not committed to reject the posits that were used to arrive at this insight.[7]

The right way to read the passage from *Natural Kinds* quoted at the beginning of this section is precisely in this immanent way: speaking from within science, Quine can say that it is revealing the structure of the cosmic machine. However, this is not to be read as the claim that science uncovers theory-independent reality.

Paragraph (3) contains a methodological reflection on the same issue. Quine writes that the semantical reflections that show that we cannot say, for instance, whether there are rabbits or cosmic complements of rabbits ("cosmic complements of" hinting at another proxy function that Quine specifies, see Quine 1995, 71) are part of the methodology of ontology and hence of (naturalized) epistemology. It is naturalized epistemology that shows that our evidence is limited in a way that does not allow us to establish whether external things exist or rather "something strange and different". Again, contemporary scientific realists would never accept this limitation of the evidence available to us. They would urge that we

[7]Another term for the same position is 'unregenerate realism', as Quine uses it in the following passage from *Five Milestones of Empiricism*: "The other negative source of naturalism is unregenerate realism, the robust state of mind of the natural scientist who has never felt any qualms beyond the negotiable uncertainties internal to science" (Quine 1981a, p. 72). Here again, Quine spells out realism in a clearly immanent way.

Caruso (2007) also examines the relationship between Quine's robust realism and contemporary scientific realism. Caruso (2007, p. 80) maintains that Quine's realism requires "taking one's conceptual scheme seriously and owning the beliefs of the moment", while insisting that the reality posited by the scheme is "constituted by our imposition of concepts, theories, and language". While his account broadly agrees with mines, Caruso never details the epistemic reasons that Quine has developed for this position.

Furthermore, this reading of Quines realism agrees with Keskinen (2012, p. 144) in its rejection of conceiving Quine as a scientific realist: "with regard to the reality of objects in some further sense than being posits of our science, Quines epistemology does not entail anything [...]".

can know that mind- and theory-independent reality contains electrons, and not their cosmic complements or, as Quine (1981b, p. 18) suggests, "classes of quadruples of numbers".[8]

In the following final passage Quine expresses his realism by resorting, ultimately, to Kantian vocabulary:

> [4] Transcendental argument, or what purports to be first philosophy, tends generally to take on rather this status of immanent epistemology insofar as I succeed in making sense of it. What evaporates is the transcendental question of the reality of the external world – the question whether or in how far our science measures up to the *Ding an sich*. (Quine 1981b, pp. 21–22)

In Quine's immanent epistemology, there is no place for the transcendental question to what extent our theory corresponds to the things in themselves, that is, to the world as it exists independently of any human cognition. Again, it is precisely such a mind- and theory-independent reality that science accurately represents according to contemporary scientific realism. And once more, the reason why Quine declares this impossible is that he subscribes to naturalism: if there is no other way of investigating reality than the scientific one, no other way to justify ontological claims than by highlighting their use in prediction, then there is indeed no epistemic ground to claim any correspondence between scientific theory and reality.

To sum up, I propose the following relationship between Quine's empiricist justificatory monism ('EJM'), his holophrastic conception of

[8] Here is another passage from *Whither Physical Objects*, which I have discussed above (Sect. 2.1.1), where Quine also suggests replacing physical objects like electrons by quadruples of numbers: "It is ironical, then, that we at length find ourselves constrained to this anti-physical sort of reduction from the side of physics itself. It is this I have wanted to bring out. Bodies were best, but they needed to be generalized to physical objects for reasons that rested on physical concerns: we wanted to provide designata for mass terms, and we wanted to accommodate physical processes or events. Physical objects, next, evaporated into space-time regions; but this was the outcome of physics itself. Finally the regions went over into pure sets; still, set theory itself was there for no other reason than the need for mathematics as an adjunct to physical theory. The bias is physical first and last, despite the airiness of the ontology" (Quine 1976, pp. 502–503).

empirical data ('HCED'), his inscrutability of reference ('IOR'), and scientific realism ('SR')[9]:

$$(\alpha)\ (\text{HCED} \wedge \text{EJM}) \Rightarrow \text{IOR}$$

According to the relationship described in (α), Quine's holophrastic conception of empirical data together with his empiricist justificatory monism implies the inscrutability of reference. If empirical evidence is holophrastic, then it does not uniquely determine which ontology to accept. Furthermore, if, as empiricist justificatory monism claims, empirical evidence is the only evidence available, then no evidence that humans could possibly get does uniquely determine which ontology accurately represents theory-independent reality. Hence, reference is inscrutable.

$$(\beta)\ \text{IOR} \Rightarrow \neg\text{SR}$$

(β) simply represents the claim that if the inscrutability of reference is true, then scientific realism must be wrong.

As a consequence, we get, by contraposition from (β):

$$\neg\neg\text{SR} \Rightarrow \neg\text{IOR}$$

That is, if scientific realism is true, then the inscrutability of reference is wrong.

And, furthermore, by contraposition from (α), we get that if the inscrutability of reference is wrong, either Quine's holophrastic conception of empirical data or his justificatory monism must be abandoned:

$$\neg\text{IOR} \Rightarrow \neg(\text{HCED} \wedge \text{EJM})$$

In total, this means that if scientific realism is true, either Quine's empiricist justificatory monism or his holophrastic conception of empirical data is wrong. Hence, rather than being a requirement of Quine's justificatory

[9]'$X \wedge Y$' represents the logical conjunction, which yields true if both conjuncts are true and false otherwise; '$\neg X$' represents the logical negation, which yields false if X is true and true otherwise; '$X \Rightarrow Y$' represents the claim that it is not possible that X is true, while Y is false.

monism or of his naturalized epistemology, scientific realism, as it is currently understood, is incompatible with it. This of course provides a very strong reason against reading Quine as a scientific realist; a reason that outweighs any remaining unease about Quine's use of 'revealed' in the passage from *Natural Kinds.*[10]

In sum, in this chapter, I have first outlined a problem for justificatory monism, what I have called the problem of justificatory pluralism. Then, I have suggested that justificatory monists can come to terms with this problem by adopting scientific realism. Finally, I have detailed contemporary scientific realism, and I have shown why Quine rejects this doctrine.

References

Ben-Menahem, Yemima. 2016. The Web and the Tree: Quine and James on the Growth of Knowledge. In *Quine and His Place in History*, 59–75. Basingstoke: Palgrave Macmillan.

Burge, Tyler. 2010. *Origins of Objectivity*. Oxford: Oxford University Press.

———. 2012. Living Wages of Sinn. *The Journal of Philosophy* 109: 40–84.

———. 2013. *Cognition Through Understanding: Philosophical Essays*, vol. 3. Oxford: Oxford University Press.

Caruso, Gregg. 2007. Realism, Naturalism, and Pragmatism: A Closer Look at the Views of Quine and Devitt. *Kriterion—Journal of Philosophy* 21: 64–83.

Castro, Eduardo. 2013. Defending the Indispensability Argument: Atoms, Infinity and the Continuum. *Journal for General Philosophy of Science / Zeitschrift für allgemeine Wissenschaftstheorie* 44: 41–61.

[10]Kemp (2016, p. 168) wants to read Quine as endorsing a species of scientific realism, as accepting that "natural science, or significant portions of it [are] literally true". While I would disagree with this claim, I follow Kemp (ibid.) insofar as this conception of scientific realism is at odds with Quine's insistence on the underdetermination of a theory by its evidence—which is a consequence of Quine's holophrastic conception of evidence. Kemp (2016, p. 170) tries to solve this tension by downplaying underdetermination. Insisting that Quine's underdetermination is to be conceived naturalistically, that is, as a thesis "whose precise content draws on our overall theory of nature" (ibid., p. 176), Kemp concludes that radical underdetermination, the one that would threaten scientific realism "looks too much like transcendental metaphysics, or even like the idle dreams of a philosopher" and hence can safely be ignored by the Quinean naturalist (ibid., p. 184). In contrast to Kemp, I would urge that Quine never accepted scientific realism, which of course resolves any tension between this doctrine and Quine's underdetermination claims. Unlike scientific realism, Quine's robust realism dovetails nicely with the underdetermination of theory by evidence.

Devitt, Michael. 1997. *Realism and Truth*. Princeton: Princeton University Press.
Egg, Matthias. 2014. *Scientific Realism in Particle Physics*. Berlin: De Gruyter.
Kemp, Gary. 2016. Underdetermination, Realism, and Transcendental Metaphysics in Quine. In *Quine and His Place in History*, ed. Frederique Janssen-Lauret and Gary Kemp, 168–188. Basingstoke: Palgrave Macmillan.
Keskinen, Antti. 2012. Quine on Objects: Realism or Anti-realism? *Theoria* 78: 128–145.
Kornblith, Hilary. 1993. *Inductive Inference and Its Natural Ground*. Cambridge: The MIT Press.
Psillos, Stathis. 1999. *Scientific Realism: How Science Tracks Truth*. London: Routledge.
———. 2009. *Knowing the Structure of Nature*. Basingstoke: Palgrave Macmillan.
Putnam, Hilary. 1990. *Realism with a Human Face*, ed. James Conant. Cambridge: Harvard University Press.
Quine, Willard Van Orman. 1969. Natural Kinds. In *Ontological Relativity and Other Essays*, 114–138. New York: Columbia University Press.
———. 1976. Whither Physical Objects? In *Boston Studies in the Philosophy of Science*, vol. XXXIX, 497–504. Dordrecht: Reidel.
———. 1981a. Five Milestones of Empiricism. In *Theories and Things*, 67–72. Cambridge: Harvard University Press.
———. 1981b. Things and Their Place in Theories. In *Theories and Things*, 1–23. Cambridge: Harvard University Press.
———. 1992. *Pursuit of Truth*, 2nd ed. Cambridge: Harvard University Press.
———. 1995. *From Stimulus to Science*. Cambridge: Harvard University Press.
Rasmussen, Joshua. 2014. *Defending the Correspondence Theory of Truth*. Cambridge: Cambridge University Press.

8

Scientifically Defending Realism I: Psillos' Holistic Approach

In the present chapter, I examine whether it is possible to scientifically defend scientific realism in the way proposed by Psillos (1999, p. 39) and Sankey (2008, p. 35). As this sort of defense proceeds on the level of entire theories—typically, it concludes that these scientific theories that are empirically successful in some specific sense of the term are true—I call this type of argument *holistic arguments.*[1]

As I have detailed above (Sect. 7.3.1), scientific realism is primarily a metaphysical doctrine concerning mind- and theory-independent reality; however, when adopting a specific kind of correspondence theory of truth, one whose correspondence relation is between claims and mind- and theory-independent facts, then scientific realism can be succinctly stated as the claim that contemporary scientific theories are typically true.

Furthermore, I have argued in Chapter 7 that justificatory monists can avoid a fatal dialectical deadlock called justificatory pluralism only if they embrace scientific realism. Therefore, this chapter and the following are dedicated to assessing whether justificatory monists can adopt scientific

[1] Usually, Putnam (1975, p. 73) is considered to have introduced this way of defending scientific realism to the modern debate; Devitt (1997, p. 109) also presents a widely discussed argument of the same sort.

R. Gubelmann, *A Science-Based Critique of Epistemological Naturalism in Quine's Tradition*, https://doi.org/10.1007/978-3-030-24524-5_8

realism. The following passage shows that Stathis Psillos embraces a version of justificatory monism:

> So, contrary to Duhem, realists argue that it is part and parcel of science and its method to rely on ampliative arguments and explanatory considerations in order to form and defend rational belief. This is exactly where the modern defence of scientific realism rests: that the kinds of argument offered by Duhem against instrumentalism can be integral to a defence of scientific realism which utilises precisely those resources and methods that scientists use to evaluate, adopt and defend their theories. (Psillos 1999, p. 37)

In this passage, Psillos recommends that scientific realists justify their doctrine using the same methods that scientists use to defend their theories. It will turn out that, according to Psillos, the main method of scientific theory justification is inference to the best explanation (IBE). Hence, by employing inference to the best explanation in his defense of realism, Psillos can claim to satisfy a version of justificatory monism. This version is characterized by its explication of empirical-scientific justification as given by IBE.

In this chapter, I first reconstruct the two so-called no-miracle arguments proposed by Stathis Psillos. Psillos' arguments are currently among the most lively debated holistic arguments (Sect. 8.1). Then, I argue that the charge of vicious circularity, one of the most common objections to such arguments, comes at a prohibitively high cost when raised against Psillos' arguments (Sect. 8.2.1). Furthermore, I argue that the so-called pessimistic meta-induction is a severe problem for Psillos' no-miracle arguments (Sect. 8.2.2). Finally, I show that Psillos' response to this meta-induction undermines his own argument, that a version of this problem also applies to Howard Sankey's no-miracle arguments (Sect. 8.2.3), and I suggest that a basic problem of this holistic way of defending scientific realism is merely accentuated through the so-called pessimistic meta-induction (Sect. 8.3).

8.1 Psillos' No-Miracle Arguments

In this section, I detail two purportedly scientific arguments for scientific realism, which are of a form that is called 'abduction' or 'IBE'. Before beginning with the discussion proper, I briefly introduce IBE as a form of inference. As Lipton (2004, p. 56) shows, this form of reasoning is very common in our everyday reasoning: given a phenomenon that is considered in need of an explanation (an *explanandum*), and given that one of the possible explanations for this phenomenon (the *explanantia*) is clearly the best, or given that it is the only conceivable explanation for this phenomenon, it is inferred that this *explanans* should be accepted. For instance, if Petra comes home and finds the door to her apartment open, even though she is certain that she has closed it when leaving, she may infer that somebody has broken into her apartment, since this is the best explanation for the scene that presents itself to her. Note that abductions are always defeasible. Unlike in the case of deductive reasoning, it is never guaranteed that the conclusion (the best *explanans*) of a sound abduction is true; all that can be expected is that it is reasonable (or, if you like, very probable) that the explanation is true.

Disagreement on the soundness of an abductive inference can arise regarding (at least) three issues. Concerning the *explanandum*, it can be debated whether there is a phenomenon in the first place and whether it is a good idea to look for an explanation of it. For example, while there is little disagreement that science is remarkably successful, empiricists like van Fraassen (1980, p. 39) have been tempted to claim that this should not be an *explanandum*. In a Kantian spirit, these empiricists could urge that the metaphysical inquiry in which the scientific realist wishes to engage is simply beyond the human intellectual powers (see also van Fraassen 2002, p. 13). Second, there can be disagreement with regard to these possible explanantia that are taken seriously. Lipton (2004, p. 149) has pointed out that, in real-world circumstances, it is never the case that all logically possible explanations are considered. Rather, the live candidates for the status of best explanation are identified by a short-list-mechanism that ensures that only those hypotheses that cohere with our background believes are considered in the first place. Thus, differences in background beliefs can lead to different shortlists and thus to different conclusions of abductive inferences. Thirdly, even if there is agreement on the *explanandum* and on the possible explanantia, there can still be disagreement on the standards used, often called the explanatory virtues, to choose the best explanation from among these possible ones. Psillos (2009, pp. 184–185) lists six such virtues: coherence with background knowledge, ability to explain *all* of the data, ability to explain *the most important* data, parsimony, unification of different explananda, and precision of prediction.

The abductive defense of scientific realism that is in the focus of this chapter has been developed by Stathis Psillos. It is generally agreed (and affirmed by Psillos [2011, p. 312]) that it proceeds in two steps, which are best conceived as constituting two distinct arguments. For reasons that will become obvious in this section, I call the first argument simply NMA (short for 'no miracle argument'), and the second argument the Meta-NMA. In a nutshell, NMA argues that the predictive success of a given theory warrants its acceptance as *correspondence true*, while Meta-NMA argues that the numerous instances of such successful and hence true theories implies that the method by which they were obtained, abduction, is typically *correspondence truth-conducive*. Psillos develops NMA it in the following passage:

> Suppose that a background theory T asserts that method M is reliable for the generation of effect X in virtue of the fact that M employs causal processes $C_1,\ldots,C_n$ which, according to T, bring about X. Suppose, also, that we follow T and other established auxiliary theories to shield the experimental set-up from factors which, if present, would interfere with some or all of the causal processes $C_1,\ldots,C_n$, thereby preventing the occurrence of effect X. Suppose, finally, that one follows M and X obtains. What else can better explain the fact that the expected (or predicted) effect X was brought about than that the theory T – which asserted the causal connections between $C_1,\ldots,C_n$ and X – has got these causal connections right, or nearly right. If this reasoning to the best explanation is cogent, then it is reasonable to accept T as approximately true, at least in those respects relevant to the theory-led prediction of X. To be more precise, more is needed for the acceptance of T as relevantly approximately true. For instance, T should be contrasted with available alternative hypotheses, and should emerge as *the* best explanation. T should also offer a 'good enough' explanation in its own right, e.g. an explanation which can adequately account for all salient features of the experimental facts. (Psillos 1999, p. 79)

I claim that the argument contained in this passage consists of the following five steps:

1. *Background theory T asserts that method M is reliable in producing effect X by certain causal processes.*
2. *Method M does produce X, when properly shielded against interferences.*
3. *T is a 'good enough' explanation in its own right. This means at least that it can accommodate all salient features of experiential outcomes.*
4. *T has been compared with alternatives and has emerged as the best explanation.*
5. *Then, the best explanation for the fact that T has correctly predicted that M does reliably produce X is that T is approximately relevantly true.*

Discussion. The most important aspect of the structure of this argument is that it contains two abductions.

On the first level, which terminates with (4), we are contrasting background theory *T* with alternatives to *T*, say *U* and *V*. According to (4), *T* has emerged as the best explanation of these three alternatives. The first question is: as the best explanation of what? Two options suggest themselves: the effect *X*, or something external to the situation described in Psillos' argument. The text is not clear at this point. However, this problem can be solved with recourse to Psillos' later elaborations of his realism and by considering (5) above. Psillos requires that to be successful, a theory has to accurately predict phenomena which were not used for the theory's construction. This means that a charitable reading requires locating the *explanandum* that is the topic of (4) outside of the situation described by NMA, since Psillos only commits himself to the truth of theories which successfully deliver novel predictions (see below, Sect. 8.2.2). Therefore, to be of any significance to (5), *X* must be a novel phenomenon; otherwise nothing would follow regarding the truth of *T* for Psillos. This means that *X* cannot be the *explanandum* in question.

Thus, we know that *T*, *U*, and *V* are competing explanations of a phenomenon other than *X*, say *Y*, and we also know that *T* has the asset of having successfully predicted the novel phenomenon *X*. Furthermore, by (4), we know that *T* has emerged as the best explanation. Psillos does not give us a rationale for *T*'s being the better explanation than its competitors *U* and *V*. In general, one would suppose that it is because *T* evinces one or more of the explanatory virtues listed by Psillos, such as coherence with background beliefs, that are not possessed by *U* and *V* (compare earlier in this section). In addition, in the present case, it may also be selected as the best explanation because these latter theories, unlike *T*, do not make any successful novel predictions. Again, this is implied by the fact that Psillos proposes to settle for the truth of *T*—if competing explanations of *Y* would also make novel predictions, *T* would probably be on equal footing with *U* and *V* with regard to the IBE of (5).

This suggests the following structure of the first IBE: scientists are confronted with a phenomenon *Y*. They devise different theories *T*, *U*, *V*, which are all 'good enough' explanations of *Y* (otherwise they would not be seriously considered). From among these explanations of *Y*, *T* possesses some explanatory virtue(s) that *U* and *V* lack. Therefore, *T* is chosen as the best explanation of *Y*.

The second IBE takes place in (5).[2] Given that *T* does successfully predict the *novel* phenomenon X, the best explanation of this prediction is that the theory in question is true.[3] Both the *explanandum* and the competing explanations in this abductive reasoning differ from the first IBE: the question is no more whether we should opt for *T*, *U*, or *V* as explanations for *Y*. Rather, the question is now whether we should adopt scientific realism, as opposed to, say, a Darwinist explanation (van Fraassen 1980, pp. 39–40) or Kant's transcendental idealism as the best explanation for *T*'s successful prediction.

To sum up, the structure of NMA turned out to be rather intricate. We are confronted with two IBE, one that selects *T* as the best explanation for a phenomenon *Y*, and another one that selects scientific realism as the best explanation for the phenomenon that *T* successfully predicts the

---

[2]To actually *draw* the abductive inference, both (4) and (5) should be supplemented with a line explicitly stating the conclusion of the inference. However, since this is not in the passage quoted, and since this does not constitute a severe problem for the argument, the reconstruction respects Psillos' actual phrasing of the argument.

[3]In fact, Psillos only claims that these theories are relevantly approximately true. For a discussion of Psillos' notion of relevant and approximate truth, see below, Sect. 8.2.2. I ignore this qualification at this point, since it only becomes relevant in Psillos' response to the so-called pessimistic meta-induction.

novel phenomenon *X*. Hence, the structure of NMA can be summarized as follows[4]:

IBE1 Explanandum. *Phenomenon Y.*

Best Explanans. *The theory T.*
Conclusion. *Therefore, T is adopted.*

IBE2 Explanandum. *The phenomenon that T successfully predicts the novel phenomenon X.*

Best Explanans. *Scientific realism with regard to T.*

Conclusion. *Therefore, scientific realism is adopted with regard to T.*

Busch (2008, pp. 108, 111) reconstructs NMA differently (while he largely agrees on the reconstruction of Meta-NMA, see Busch 2008, p. 108). He does find the second IBE, as it is part of the reading here, but he considers what is here read as the first IBE not to "fit the model of abduction" (Busch 2008, p. 111). The reading presented here does analyze the relevant part of Psillos' reasoning as an abduction, but it had to make an effort to do so. Thus, it is understandable that Busch does not find any abductive inference on the first level—the text is simply not entirely clear on this. Still, reading the first IBE as an abduction is the more charitable reading, given Psillos' comments on the reasoning (he calls it "a more concrete type of explanatory reasoning" on Psillos [1999, p. 79]), and the importance of its being an abduction for his overall strategy.

I now move on to the second part of Psillos' purportedly scientific defense of scientific realism, the Meta-NMA. In the following passage, Psillos presents the argument.

> But how exactly does this argument defend IBE and hence how exactly does NMA become the pivot for a realist epistemology of science? As I

[4]The schema used in this overview largely coincides with the proposal by Josephson (1994, p. 5), which is also employed by Busch (2008) and by Psillos (2007, pp. 442–443). What is called *explanandum* here is referred to as a set of data (facts, observations, givens) by Josephson (ibid.). The *explanans* is Josephson's hypothesis that would explain the data, if it were true. The emphasis on the need for the *explanans* to be the best explanation of course raises a number of issues. In particular, if the best is taken in an epistemic way, that is, to be read as: the best explanation available to the researchers of a given time and place, it raises the problem of unconceived, but better alternative explanations, as suggested by Stanford (2006).

> have noted, it suggests that the best explanation of the instrumental reliability of scientific methodology is that background theories are relevantly approximately true. These background theories have themselves been typically arrived at by abductive reasoning. Hence, it is reasonable to believe that abductive reasoning is reliable: it tends to generate approximately true theories. (Psillos 1999, pp. 79–80)

I claim that it can be reconstrued as follows:

Explanandum. *The relevant background theories of numerous instrumentally successful scientific methods are correspondence true.*

Best Explanans. *The method by which these background theories have been developed, i.e. abduction, tends to generate correspondence true theories.*

Conclusion. *Therefore, abduction tends to generate correspondence true theories.*

Discussion. The *explanandum* is given by the outcome of the argument on the first level: the theories which have made successful predictions and which satisfy some other requirements are true, and it is claimed that these theories have been arrived at by abductive reasoning. As the best *explanans* for this phenomenon we are then offered the hypothesis that the method by which they have been obtained, that is, IBE, tends to generate true theories. By IBE, it is then concluded that IBE tends to generate true theories.

This analysis disagrees with the one by Iranzo (2008). He suggests that the second subargument is a *deductive*, instead of an abductive inference. This reading is inconsistent with the explicit comments by Psillos (1999, p. 79) on his argument: he calls it a meta-abduction. Furthermore, Iranzo is forced to include a premise not explicit in Psillos' wording: "An inference rule is reliable iff it yields a high rate of approximately true conclusions" (Iranzo 2008, p. 116). Remarkably, the only reason given by Iranzo for the inclusion of this additional premise is that it would not be deductively valid without it. However, it seems misguided to think that Meta-NMA is intended as a deductively valid argument. Rather, it is more in the spirit of Psillos' approach, and consistent with his explicit comments on Meta-NMA, to put the presupposition as follows—if one wants to include it at all: "If we have observed in the past that an inference rule generates a high rate of true conclusions, then this observation is best explained by holding that it does so in general, that is, that it typically generates true conclusions." This, however, keeps the abductive nature of the argument intact.

The conclusion of Meta-NMA corroborates a realist epistemology of science since, according to Psillos (1999, p. 37), scientific inquiry employs IBE to develop and to justify its theories. In addition to establishing that the main method of science tends to generate true theories, Meta-NMA implies much more: if successful, it establishes that abduction simpliciter is truth-conducive.

Therefore, the conclusion applies to the argument itself: if it is true that IBE tends to generate true conclusions, then this would justify Meta-NMA itself as well as NMA, being IBE. This self-referential structure is not only self-stabilizing, it also provokes the most prominent critique of NMAs, according to which they are viciously circular (see below, Sect. 8.2.1).

To sum up, Psillos proposes two arguments for scientific realism. The first one, NMA, aims to establish the correspondence truth of specific scientific theories: it urges that the best explanation for scientific theories' successfully predicting novel phenomena is that their ontological posits are instantiated in theory-independent reality. Thus, scientific realism with regard to these theories should be adopted. The argument contains two abductions: one of them selects a certain scientific theory as the best explanation for a given natural phenomenon, the other selects scientific realism as the best explanation for said scientific theory's novel predictive success. The second argument is based on specific instances where, according to the first argument, theories turned out to be correspondence true, and it infers from this that abduction in general yields correspondence true theories.[5]

[5]Compare Fitzpatrick (2013, p. 149) for a very similar reading of the second IBE of NMA and of Meta-NMA.

8.2 A Threefold Critique of Psillos' No-Miracle Arguments

In this section, I develop a threefold critique of the two arguments for scientific realism that I have reconstructed in the preceding section. First, I argue that the arguments are not viciously circular—unless one is prepared to accept a global skepticism regarding any inferences whatsoever. However, I also argue that the same property that immunizes the arguments against the charge of vicious circularity also implies the failure of Meta-NMA (Sect. 8.2.1). Then, I detail the so-called pessimistic meta-induction and Psillos' response to it, and I suggest, somewhat tentatively, that Psillos' response is unsatisfactory (Sect. 8.2.2). Finally, I show that this same response undermines the credibility of NMA itself (Sect. 8.2.3).

8.2.1 On Vicious Circularity—And on Why Meta-NMA Fails

The objection of vicious circularity against no-miracle arguments has been put forward in particular by Fine (1991, p. 82). Psillos accepts this objection as "the main line of criticism" against NMA (Psillos 1999, p. 81). Lipton (2004, p. 186) is convinced that this circularity-objection destroys the dialectical force that Psillos wants his arguments to have: "The circularity objection does I think show that the miracle argument has no force for the non-realist." Put briefly, the concern here is that neither of the two arguments has any dialectic force because they presuppose precisely what is disputed by their target audience, namely the claim that abductions yield true conclusions.

Before addressing this critique, I wish to point to an important dialectical effect of this challenge: opponents of NMA that level this charge *accept that the argument employs the method of science to justify its hypotheses.* The circularity charge cannot possibly be put forward if one holds that NMA employs a different method of justification than the method of the sciences, whose truth-conduciveness is at issue. Thus, the opponent to NMA has a dialectic price to pay if she wishes to argue that the argument is viciously circular. She has to grant that NMA is a scientific argument, thus confirming the claim of Psillos and other justificatory monists, according to which they employ the same way of justifying claims that is also used by scientific inquiry.

Now, what precisely does the criticism amount to? According to Fine (1991, p. 82), the problem is as follows. At issue is the relationship between explanatory success and truth. According to the realist, the explanatory success of a given theory implies that it should be accepted as true. According to the non-realist, however, there is no such implication. Now, Psillos the realist presents a reasoning to the conclusion that predictively successful scientific theories are by and large true, and another reasoning to the conclusion that abduction tends to generate true theories. Both of these arguments are abductions. To this, Fine would respond that the arguments do not advance the discussion, since they presuppose precisely what they aim to establish, namely the truth-conduciveness of abduction: non-realists will not be convinced by these arguments since they are non-realists precisely because they doubt that abduction is truth-conducive in general.

Psillos responds to this charge using two different and independent strategies. The first one consists in encouraging an externalist stance towards epistemology. The most important benefit, which he expects from this move, is that he is permitted to use a rule of inference without being able to justify it because externalists focus exclusively on the reliability of a given form of inference, and not on the user's ability to justify it. Regarding the question at issue presently, this means that the externalist is justified to use abduction even though he himself cannot justify this use, as long as "the rule of inference is reliable (this being an objective property of the rule)" (Psillos 1999, p. 83). In this sense, the sensor of an automatic door can be said to be justified in its judgment that a body of some extension is approaching, since its rule of inference—something like: given a certain pattern of infrared light reflection, assume that a person is approaching (and hence open the door)—is reliable. However, this does not seem to help in the current debate. As it were, non-realists are no automatic doors: they are non-realists precisely because they see a lack of justification in the realist's claims that scientific theories are typically true and that abduction is typically truth-conducive. By adopting the externalist's stance, Psillos thus scares away his target audience.[6]

While the prospects of Psillos' externalist strategy are rather dim, his second approach to counter the circularity-charge is successful. This approach appeals to the fundamentality of abduction as a rule of inference, and it employs it in two ways against the circularity-charge leveled by the non-realist. Firstly, he points out that abductive skepticism cannot be maintained in one's daily business (Psillos 1999, p. 89). Due to the omnipresence of abduction, there seem to be countless circumstances in which any rational being would draw an abductive inference and naturally assume that it is truth-conducive. Thus, Psillos (1999, p. 211) claims that the "patterns of reasoning, as well as the justification" are the same in cases of commonsense abductive reasoning and of scientific IBEs. To take a homely example: if such a proponent of the circularity-charge comes home and finds the ice-cream compartment of the fridge empty, she may conclude that it was her son Pete who is responsible for it if she knows that Pete has a habit of eating huge amounts of ice cream occasionally and that all other members of the family are away on vacation. Very likely, Pete will not get away with pointing out that his mother is an abductive skeptic and therefore cannot infer anything from the fact that Pete's having eaten the ice-cream is the best explanation for the empty ice-cream compartment. Very likely, the skeptic will not choose this option, but rather specify her skepticism. She could restrict her skepticism to the domain of unobservable, theoretical entities, which would save her from the awkward duty to doubt any abductive inferences whatsoever. However, by the same token, the charge of vicious circularity would evaporate: the problem to be discussed then would no more be the form of abductive inferences, but certain epistemological reservations against the positing of unobservable entities in explanations.

In addition to this, Psillos (1999, pp. 86–89) correctly points out that abduction fares no worse than deduction in terms of its justification. To argue for this, Psillos first introduces a distinction between premise- and rule-circularity. An argument is premise-circular if its conclusion figures

[6]Additionally, Newman (2010) has argued that such an externalist reading of NMA faces the so-called Generality-Problem, which occurs in many reliabilist contexts.

as one of its premises. Such circularity is admittedly vicious. In contrast, rule-circularity is deemed non-vicious. An argument is rule-circular if it infers, by the use of inference-rule *R*, the reliability of *R* (see Psillos 1999, p. 82 and Papineau 1993, pp. 157–158). Psillos (2009, p. xvii) rightly claims that his NMAs are not premise- but only rule-circular. Furthermore, he holds that "rule-circularity is involved in the defense of all basic and fundamental rules of inference" (see also Psillos 1999, p. 89). This is correct, and it is one possible moral that can be drawn from Lewis Carroll's ingenious *What the Tortoise Said to Achilles* (Carroll 1895): any attempt to justify *modus ponens* by *modus ponens* will end in a regress. This means that those who refuse to use abduction because any justification of abduction is itself abductive must, by parity of reasoning, refuse to use deduction as well.[7]

In summary, one of the most important critiques of this kind of defense of scientific realism, the charge of vicious circularity, was shown to be misguided. Any critic of said defense who employs this charge is not only committed to accept that the defense does indeed employ the scientific method of justification; she is also committed to skepticism with regard to abductions in everyday circumstances and, indeed, by parity of reasoning, with regard to deductive inferences. In other words, leveling the charge of vicious circularity implies a global skepticism with regard to any inferences whatsoever. Since few philosophers would be prepared to accept this kind of skepticism, the charge of vicious circularity will not greatly diminish the possible target audience for both NMA and Meta-NMA.

* * *

After having shown that the charge of vicious circularity, leveled against both NMA and Meta-NMA comes at an unaffordably high price, I now focus on a severe problem of Meta-NMA.

Unfortunately, *the very same generality of abductive reasoning that saves Psillos' NMA from the charge of vicious circularity also implies the failure of Meta-NMA*. To see the problem with Meta-NMA, consider the following

[7] In contrast, in the dialectic situation delineated above (Sect. 7.1), the target audience of a defense of *J1*, say proponents of *J2*, can avoid any use of *J1*. Hence, this target audience can level the charge of vicious circularity against a defense of *J1* that itself observes *J1* at no such cost.

scenario: a tribe on a remote island in the pacific has come to believe that tempests are caused by angry spirits. Their justification for this belief proceeds by IBE: they know that spirits are very powerful and also very cruel when angered. Furthermore, they know that no fellow tribesman or -woman is powerful enough to cause such a remarkable phenomenon. Therefore, the best explanation of tempests is that they are caused by angry spirits. This reasoning is abductively sound (to the extent to which an abduction may be said to be sound): there is no doubt that their *explanandum* is real: tempests are real, and there seems to be no reason not to try and explain them. Furthermore, the tribe is presented with two alternative explanations for the tempest, and its members choose the best of these two based on the criterion of power: humans cannot cause such a phenomenon, but spirits can. Still, this explanation fundamentally contradicts modern physics.

Of course, meteorologists could engage members of the tribe in a debate and try to convince them that modern physics has a better explanation for tempests, one that also makes reference to unobservable entities, but not to minded ones. To this, a sophisticated member of the tribe could reply that a good explanation should be simple and intuitive, and that according to this standard, the tribal explanation is superior to the one of the meteorologist. Presumably, the meteorologist would maintain that her standard is predictive success: her theory, even though counterintuitive and rather complex, is able to predict the weather up to five days in advance.

Since the conclusion of Meta-NMA is not restricted to abductions that are based, say, on the explanatory virtue of predictive success, but rather maintains that, regardless of any further specification, abduction is truth-conducive, this scenario illustrates a serious problem for the argument: both the tribal and the scientific abduction are sound and should therefore be true, according to Meta-NMA.

One can say that the problem in this example is that the tribesman and the meteorologist have very different ideas of what it takes to be an explanatory virtue, that is, what it takes to be the best explanation for a given phenomenon. Harman (1965, p. 89) mentions this problem already in his seminal article *The Inference to the Best Explanation.*

Alternatively, one can explain the disagreement between the tribesman and the meteorologist by pointing out that they have completely different background beliefs. As mentioned above, Lipton (2004, p. 149) suggests that the list of possible candidates for the position of best explanation is determined by the relevant background beliefs. In our case, it is very plausible that what is the best explanation for the meteorologist would not even make it on the tribesman's shortlist, and vice versa. Psillos (2009, p. 188) himself emphasizes the role of background beliefs by pointing out that coherence with background beliefs is central for the justification of abductions: the "coherence-enhancing role of IBE [...] is ultimately the warrant-conferring element of IBE." Given a case where background beliefs differ as starkly as between the meteorologist and the tribesman, it is very likely that the relevant abductions will confer warrant to obviously inconsistent conclusions.[8] Fitzpatrick (2013, p. 149) and indeed Psillos (2007, p. 443) have emphasized the context-sensitivity, or background-dependence of IBE. This context-dependence is also at the root of the problem delineated in this section: whether a given abduction is valid depends crucially on the background believes involved.

Could this importance of background beliefs for the abductive disagreement between the meteorologist and the tribesman be used to resolve the problem rather than merely to explain it? After all, Lipton (2004, pp. 149–151) argues, using a Darwinian analogy, that the items constituting our background knowledge do constitute our background knowledge because they were once chosen as best explanations. However, this only postpones the problem, since it cannot explain the emergence of, as it were, the first set of background beliefs themselves: they cannot be the product of abductive reasoning, since abductions always presuppose background beliefs. *A fortiori*, it will not help to solve the problem of radically different background beliefs.

The central moral to be drawn from this analysis is that the conclusion of Meta-NMA is too broad to be true, even if Meta-NMA's *explanandum* is conceded: it is simply wrong that the mere abductive form of a reasoning justifies belief in the truth of the conclusion.

[8] A very similar observation leads Stanford (2006, pp. 27–37) to raise the problem of unconceived alternatives. Stanford argues that the abductive argument for scientific realism would be valid only if scientists considered all possible explanations for a given explanandum. Since they cannot possibly do that, the argument is not valid.

8.2.2 Psillos' Response to Historical Counterexamples to NMA

After having argued that the conclusion of Meta-NMA is false because it allows wildly inconsistent hypotheses to be true, I now focus on NMA. The basic inference of this argument is the one from predictive success to correspondence truth. In this section, I consider the most important challenge to this inference, and I detail Psillos' response to it, which will include a delineation of his notion of novel predictive success.

The core of the historical challenge is that there were scientific theories that were clearly successful but still, judged by contemporary standards, fundamentally wrong. Laudan (1981, p. 33), who is credited with introducing the issue into the contemporary debate, presents an impressively long list of such theories, among them the caloric theory of heat, the phlogiston theory of chemistry, and theories of spontaneous generation. According to Laudan, all of these theories were not only endorsed by the relevant expert community, but are also "successful by any criterion [he] can think of (including fertility, intuitively high confirmation, successful prediction, etc.)." However, according to current scientific theories, essential parts of these theories are wrong. For the realist, this stark change of ontological commitment of chemistry is a possible source of trouble because he wants to learn about mind- and theory-independent reality from scientific theories, and about its unobservable, fundamental structure in particular. Modern chemistry and phlogiston theory give two different descriptions of this fundamental structure—if you want to know what the fundamental structure of the world is, you have to choose.

In fact, our contemporary theories, say modern chemistry, do rarely strictly contradict ontological claims of former theories, say phlogiston theory: Contemporary chemistry does not contain the claim: phlogiston does not exist.[9] Rather, there is no need, indeed no room in contemporary chemistry for phlogiston; when considered from the point of view of modern chemistry, positing phlogiston is not only explanatory superfluous, it makes no sense: phlogiston was postulated as a constituent of metals and other materials that is necessary for combustion processes and is consumed

[9]This point was urged by Marco Toscano.

during the process. Hence, materials that leave little ash were supposed to be rich in phlogiston (compare Weisberg et al. 2011 for details). For contemporary chemistry, positing phlogiston makes no sense because the process of combustion is explained in fundamentally different ways, as a kind of oxidation process. As a consequence, the scientific realist needs a reason that is not ad hoc that justifies his acceptance of contemporary chemistry's posits and his rejection of the posits of phlogiston theory.

Note that a successful answer to this historical challenge does not provide a positive reason for the doctrine, it merely removes a defeater for it. Only after removing this defeater may the naturalistic realist hope that NMA goes through, thus giving her a positive scientific reason for realism, showing that science can address the realism-antirealism dispute.

Psillos (1999, pp. 101–104) conceives of this challenge as being inductive in form, namely as constituting a pessimistic meta-induction: given that so many successful past theories turned out to be wrong in important ontological aspects, it seems that it can be inductively inferred that our current theories are going to meet the same fate. Lyons (2002, p. 65), however, argues that Laudan's argument's logical form should be conceived as a *modus tollens* rather than an induction. The realist argues that the astonishing success of our scientific theories would be a brute miracle if they would not accurately represent theory-independent nature. Since assuming miracles is an abductive catastrophe, the argument goes, we are justified to hold that any successful theory is correspondence true. In other words, the realist presents us with a *modus ponens* of the following form:

P1 If a theory is successful, it accurately depicts theory-independent, ahistorical nature (otherwise its success would be a miracle).

P2 Our contemporary scientific theories are successful.

K Hence, our contemporary scientific theories accurately depict theory-independent, ahistorical nature.

As mentioned above, (P1) is supported by the so-called no-miracles consideration: we are supposed to accept (P1) on pain of having to believe in brute miracles.

Against this, Laudan's list is intended as a *modus tollens* of the following form (quoted from Lyons 2002, p. 65):

P1 If the realist hypothesis is correct (A), then each successful theory will be true (B).
P2 We have a list of successful theories that are not true (Not B).
K Therefore, the realist hypothesis is false (Not A).

On this reading, Laudan's argument is not an inductive reasoning from the fact that many past theories were wrong to the conclusion that our present theories are probably wrong, too, but rather a *modus tollens*, that is, an argument for the falsity of the realist's hypothesis, which claims in its simplest form that "theories enjoying general predictive success are true" (Lyons 2002, p. 63). Thus, regardless of the no-miracles consideration, one should not accept the realist's hypothesis.

Lyons is right to suggest that Laudan intended his list of historical examples not as a basis for an inductive inference to the falsity of our present theories, but rather as a challenge to a general claim made by realists. This follows from systematic considerations. Both Lyons (ibid.) and Psillos (1999, p. 105) rightly remark that Laudan's list would be a rather thin basis for any inductive inference. Hence, charity requires that the list be not read as a basis for an inductive inference.

While I follow Lyons in his reading of Laudan's historical challenge this far, it seems unnecessary to conceive of the argument as a *modus tollens.* Consider again *P1* in Lyon's *modus tollens* above, but bear in mind that the realist hypothesis claims that success implies correspondence truth. Then premise one becomes quite tautological: if it is correct that success implies truth, then successful theories are true. Instead, I propose to read Laudan's argument as having the simple form of giving counterexamples to a general claim made by the scientific realist. The scientific realist claims that, for any theory *T*, if *T* is successful (in the respective realist's preferred sense of 'success'), then it is true. Otherwise its success would be a miracle. Against this, Laudan lists examples that contradict this general rule and thereby falsify it. In what follows, I therefore call objections against realism that follow this general pattern 'Laudan-Style Counterexamples', or 'LSC'.

* * *

Psillos' response to Laudan-Style Counterexamples (LSC) has been called "[t]he most detailed an influential response to the argument from theory change", see Ladyman (2014). The response has three parts. First, Psillos makes clear that he is willing to commit himself only to the correspondence truth of scientific theories that deliver accurate predictions of novel phenomena, so-called "novel predictions" (Psillos 1999, pp. 35, 76, 105). For example, Psillos mentions that the "[t]idal phenomena were predicted by Newton's theory, but they were not used in its construction" (Psillos 1999, p. 106). More generally speaking, Psillos requires that to be successful, a theory has to accurately predict phenomena that were not used for the theory's construction. This explication of novelty allows for predictions to be novel even if the respective phenomena have been known at the time of the construction of the theory, as long as information about the predicted phenomena has not been used for the construction of the respective theory.

Secondly, Psillos (1999, p. 105) maintains that only mature theories are true. According to Psillos, theories are mature if "they [are] representative of their disciplines at stages of development sufficiently advanced as to be reckoned theoretically mature." Thus, theoretical maturity is a property of the discipline to which the theory belongs. Here is how Psillos elaborates on his criterion of maturity:

> Realists require that Laudan's list should include only *mature* theories; that is, theories which have passed the 'take-off point' (Boyd) of a specific discipline. This 'take-off point' can be characterized by the presence of a body of well-entrenched background beliefs about the domain of inquiry which, in effect, delineate the boundaries of that domain, inform theoretical research and constrain the proposal of theories and hypotheses. This corpus of believes gives a broad identity to the discipline by being, normally, the common ground that rival theories of the phenomena under investigation share. It is an empirical matter to find out when a discipline reaches the 'take-off

point', but for most disciplines there is such point (or rather a period). (Psillos 1999, pp. 107–108)

Note that this is a sociological criterion: it depends on the homogeneity of professed background beliefs of the relevant community whether a discipline is mature or not.

The third part of Psillos' response to LSC consists in weakening the conclusion: Psillos (1999, pp. 108–112) does not claim that the mature theories which successfully predict novel phenomena are true, but only *approximately relevantly* true.[10] Essentially, the idea is to offer the LSC as small an attack surface as possible. By only claiming *approximate* truth of (mature and predictively successful) theories, the no miracle argument is immunized against small changes in theory (where it is of course a difficult task to precisely specify what 'small' means here). By only claiming truth of *relevant* parts of the theory, that is, of those constituents which are actually responsible for the success of the theory, the argument is immunized against changes of parts of a theory which are not involved in the success.

In sum, Psillos applies a threefold criterion, or three filters to identify those theories to whose correspondence truth he commits himself. Two of these filters operate on entire theories, namely the criteria of maturity and of novel predictive success, while the third, relevance, applies within specific theories that have passed through the first two filters. Finally, Psillos is willing to commit himself only to the approximate truth of the remaining bits of theories.

I suggest that there are substantial problems with Psillos' response to LSC as such. First, with regard to maturity, Laudan (1981, p. 34) argues that this criterion is not convincing, since it makes the refutation of LSC circular by only counting those theories as mature which are sufficiently similar to our current theories, and since it leaves the success of all immature theories unexplained, that is, miraculous. Furthermore, truly revolutionary proposals like Newton's in physics, that basically redefined the entire fabric of background beliefs in physics, could not be taken to be

[10] As nothing in my discussion of Psillos' arguments depends on this qualification, I do not make this restriction explicit in the following—it should be understood, however, that Psillos always only claims approximate and relevant correspondence truth of the theory in question.

correspondence true, since physics, by the time of Newton's proposal, was not a mature discipline by Psillos' standards. Finally, there is almost complete agreement among the practitioners of so-called creation science with regard to their background beliefs (a rather naive reading of the book Genesis of the bible), but Psillos would probably not want to call this discipline mature, and to count this sociological feature of this community as speaking in favor of the correspondence truth of their theories.

Second, with regard to novel predictive success, Lyons (2002) presents an impressively long and variegated list of theories to which the realist most likely does not want to commit herself, but which were still successful in delivering novel predictions. So, maturity and novel predictive success do not seem to constitute attractive sufficient conditions for the realist's choice of theory. Furthermore, Lyons (2009, pp. 77–78) argues that several of the theories to which the scientific realist most likely wants to commit herself have not excelled through novel predictions, among them Newton's theory of universal gravity. More generally speaking, he argues that *any* description of trivial facts that is in principle unable to make any novel predictions is trivially true, thus questioning whether novelty is even a *necessary* condition for truth.

To reply to this challenge, the realist could try to show that the items on the list are, *pace* Lyons, not real instances of theories that are successful in delivering novel predictions. The realist could also modify her criterion such that the items on the list do no longer fall under that criterion. Finally, she could argue that the theories are not really committed to the entities which are most problematic, like phlogiston, the optical ether, or the caloric. Indeed, Psillos (1999, pp. 110–139) does precisely that for the caloric and the optical ether. It seems much more difficult, however, to argue along the same lines in the case of phlogiston. Phlogiston seems to be irreducibly involved in novel predictions. Thus, Lyons (2002, p. 70) refers to a (successful) novel prediction by Stahl, according to which sulfuric acid and phlogiston, which is emitted from glowing coal, combine to create sulfur. This agrees with the analysis of Ladyman (2011), who argues that the case of phlogiston undermines scientific realism understood as a kind of entity realism and strongly suggests a structural realism.

More generally speaking, the way this historical question is debated at present is analogous to a *race of arms*: the scientific realists propose a reading of success, then non-realists come up with some supposed counterexample. If it turns out that the example in question does indeed satisfy the respective criteria for success, while still being, from a contemporary point of view, misguided, the realists return to their studies to refine their notion of success such that the newly proposed example is excluded. Arguably, the non-realists currently have the upper hand in this race of arms. Furthermore, as historians of science continue to broaden and deepen their accounts, the task becomes more difficult for the realist because the account of the history of science becomes more multi-faceted and diverse. As it were, there is an ever-increasing chance that some historian comes up with a theory of medieval alchemy that was successful in the latest sense, but still wildly incompatible with contemporary theories.[11]

To conclude, I highlight again the dialectic potential of the LSC debate. Let's assume that, contrary to current trends, the realist succeeds in specifying a notion of success that holds water against any LSC whatsoever. Does this imply that the entities postulated by (parts of) theories that are successful in this sense do exist in mind-and theory independent reality? No. It merely removes a defeater for this claim, namely the fact that the supposed criterion commits its defenders to a multitude of inconsistent ontological posits (in the sense of 'inconsistent' elaborated at the beginning of this section). An LSC-immune conception of success could therefore remove a defeater for somebody who finds realism intuitively

[11]As many arms races bring with them, as a side effect, technological progress and substantial scientific insight, the arms race between proponents of NMA and anti-realists has produced significant insight into the history of science as a side effect. The history of science, whose relevance for scientific realism was evinced by the pioneering work of Kuhn (1996 [1962]), has flourished as a consequence of the boldly realist claims embodied in NMA. The main insight provided by this race of arms is that science, conceived as a historical phenomenon, is extremely multi-faceted and diverse. As a consequence, it becomes very difficult to maintain a simple narrative of linear progress. Put slightly polemically, this narrative goes as follows for the science of physics. Isaac Newton invented modern physics. There was no serious physics before Newton. His mechanics carved nature almost at its joints. Subsequent developments in physics—in particular Einstein's theory of general relativity—mostly just improved on the accuracy and the scope of Newton's mechanics. According to this narrative, as science progresses, its theories depict the true forms of nature ever more accurately, improving on the preceding theories, but not overthrowing their ontological claims at root.

The debate of the simple inference at the core of NMA—that we are justified to assume that empirically successful scientific theories are mostly accurate descriptions of the world itself—has produced a wealth of historical evidence that evinces this narrative as much too simplistic (and arguably as entirely misguided).

plausible, but who has some scientific self-doubts[12]—but it cannot constitute a positive scientific argument for scientific realism.

8.2.3 Psillos' Response to Historical Counterexamples Undermines NMA

In the preceding section, I have introduced the challenge of LSC to Psillos' NMA, and I have suggested that the realist may lose this conceptual race of arms against the nonrealist as our account of the history of science becomes more nuanced and detailed. Furthermore, I have pointed out that a successful response to LSC does not constitute a positive argument for scientific realism.

In this present section, I suggest that the same response to LSC implies that scientific realism is unjustified, which means that the naturalist was never entitled to assume the realism she wants to defend with LSC in the first place: the second IBE of NMA is in no way recognizably empirical, thus falling clearly short of Psillos' own strict standard for the justification of a scientific theory, namely novel predictive success. I suggest that the basic reason for this problem is Psillos' justificatory monism: since there is only one standard by which to decide whether a theory is justified, it follows that the very strict standard of justification that Psillos develops for scientific theories in response to LSC must also be applied to scientific realism. Hence, Psillos cannot answer LSC, maintain justificatory monism, and defend scientific realism all at once. To conclude, I point out that a similar defense of scientific realism, due to Howard Sankey, faces the same problem.

As shown above (Sect. 8.1), the central inference of NMA is from the novel predictive success of a given theory T to the adoption of scientific realism with regard to this theory. For proponents of Psillos' response to LSC who are also justificatory monists, the basic problem with this inference is that scientific realism clearly doesn't make any novel predictions even remotely similar to those cases listed by Psillos (see above, Sect. 8.2.2). Since justificatory monists apply the same justificatory standards to

[12]The expression 'scientific self-doubts' is due to Quine (1974, p. 3).

scientific as well as to metaphysical theories, this means that scientific realism itself has to be novel predictively successful. However, scientific realism clearly does not deliver any such predictions. Hence, it falls clearly short of Psillos' self-set standards for the correspondence truth of a given theory. This means that scientific realism should not be adopted with regard to any theory, according to the criteria that Psillos himself develops in his response to LSC. The problem can be put in the form of the following dilemma:

Horn 1. Introduce new, non-empirical standards that a *philosophical* theory has to fulfill to be accepted as correspondence true. These standards differ from the ones by which scientific theories are measured. This dualism of justificatory standards, however, would decisively contradict Psillos' justificatory monism.

Horn 2. Stick to justificatory monism. This forbids Psillos to consider scientific realism as correspondence true, as it is neither successful nor mature in the relevant sense. More precisely, embracing this horn forbids Psillos to draw the second IBE of NMA, which means that the scientific theory *T* simply emerges as the one that has surpassed all of its competitors in terms of predictive success, but not as one whose ontological posits accurately represent mind- and theory-independent reality.

To conclude this critique of Psillos' NMA, note that there is a stark tension between Psillos' response to LSC and the conclusion of Meta-NMA. According to the latter, we should accept as correspondence true all theories that are generated by abduction. According to Psillos' response to LSC, however, we should accept as correspondence true only these theories that successfully deliver novel predictions and that stem from mature disciplines. As the case of the tribal explanation for tempests evinces, these two delimitations of true theories are not coextensive, as the tribal explanation would count as warranted according to the conclusion of Meta-NMA, while it would not come out as justified when measured by the standards of Psillos' response to LSC.

The tension between these two ways of carving out these theories to whose correspondence truth Psillos' commits himself should give one pause. I suggest that the reasons for this double standard are intrinsic to the very project of a scientific defense of scientific realism. On the one hand, a proponent of such a project is pressured to conceive the threshold of justification permissively enough to count his own defense not only as scientific, but as belonging to this subclass of scientific theories that are true; hence the extremely generous standard of Meta-NMA. On the other hand, the same proponent is out to defend a position that is incompatible with ontological pluralism: there is only one way the world is, and the sum of all theories he commits himself to had better depicted a uniform and consistent picture of this way the world is. Thus, in light of LSC, Psillos is pressured to be very strict when it comes to delimiting the class of theories that he wants to count as justified and hence correspondence true.

In the following, I show that a similar attempt at a scientific argument for scientific realism, due to Howard Sankey, also contains such an untenable double standard of justification. Sankey (2008, pp. 112–113) first affirms that there is no a priori connection between truth and rational justification, since the former is to be understood non-epistemically as correspondence to mind-independent states of affairs. As a consequence, a gap between scientific method of justifications and truth opens, and the abductive defense of scientific realism comes to the fore: "In the absence of direct or conclusive evidence linking method of justification to truth, the grounds for such a link may be at best abductive ones" (Sankey 2008, p. 118). Sankey then invites us to consider an example of a theory which is successful in every conceivable sense, which he thinks is tantamount to *satisfying the rules of scientific method* (in particular, the rules that concern the selection of true theories). Then, he argues that the best explanation of the success of this theory is that it is approximately true. This reasoning is strikingly analogous to the one of Psillos' NMA. Furthermore, in analogy to Psillos' abductive defense, Sankey employs this conclusion to argue for the truth-conduciveness of scientific method of justification in general:

> Given the critical role played by the rules of method in the process of theory selection, the implications of the success of science for the approximate truth of theory apply with equal force to the rules of method themselves. [...] That is, the rules of method are rules that 'screen for truth' – or, in other words, the rules of method reliably conduce to truth. (Sankey 2008, p. 119)

The analogy between this reasoning and Meta-NMA is striking. Sankey argues that the truth of theories is best explained by the truth-conduciveness of the so-called rules of method.

Depending on the specific content that Sankey assigns to 'the rules of method', his argument will fail in the manner of Meta-NMA or of NMA: if he takes only the mere form of inference, abduction, to constitute 'the rules of method', then his argument will meet the fate of Meta-NMA. Such a broad notion is suggested by the manner in which he defends the claim that his theorizing is analogous to scientific research:

> For to treat realism as a hypothesis about science, and to argue for the hypothesis of realism by means of inference to the best explanation, is to treat realism as a hypothesis that is to be evaluated in a manner analogous to the evaluation of scientific hypotheses. (Sankey 2008, p. 35)

This passage, where Sankey argues that the method of justification that he uses to defend scientific realism is the scientific method of justification, employs the same unqualified conception of abduction as in Meta-NMA. Thus, if in the first passage from Sankey quoted above, "rules of method" is to be read in this way, then the argument fails in the very same way as Meta-NMA.

There is, however, another much more demanding notion of the rules of method at work in Sankey's study of scientific realism. Such a richer notion of rules of method is suggested when Sankey specifies what he means by a successful theory to whose truth he is willing to commit himself:

> Suppose, for example, that there is some theory which satisfies a broad range of rules of method to an extraordinarily high degree. The theory is supported by all available evidence. It successfully predicts a great many previously unknown and surprising novel facts. It unifies previously disparate domains. And it does all of this in a manner which maximizes simplicity and coherence. Clearly, any theory which so impressively satisfies the rules of scientific method is a highly successful theory indeed. [...] Where a theory impressively satisfies a broad range of methodological rules, the best explanation of such success is that the theory provides an approximately true description of the way the world is. (Sankey 2008, p. 118)

Even though Sankey never delivers a clear-cut list of those methodological rules whose satisfaction is necessary and sufficient to make commitment to the respective theory's truth rational, it is clear from considering what he lists in this passage that scientific realism cannot possibly be successful in this sense, hence it should not be adopted with regard to any scientific theory. In short, we find that Sankey's defense of scientific realism is not only analogous in structure to Psillos' NMA and Meta-NMA, it also fails for reasons that are analogous to the reasons for the failure of Psillos' arguments.

Indeed, this insight arguably generalizes. It may present a general difficulty for justificatory monists that try to defend scientific realism following the holistic paradigm instantiated by NMA, that is, that argue that scientific theories satisfying a certain criterion of empirical success should be taken to accurately represent mind- and theory-independent reality (or to do so approximately and relevantly). It may not be possible to develop a conception of empirical success that is both strict enough to withstand any Laudan-Style Counterexamples and generous enough to evince scientific realism itself as justified with regard to any scientific theory.

8.3 Provisional Conclusion: Empiricist Justificatory Monists Cannot Defend Scientific Realism

In this chapter, I have critically assessed the two arguments Psillos gives for scientific realism (I have called these arguments 'NMA' and 'Meta-NMA' respectively). After pointing out that leveling the charge of vicious circularity against any of the two arguments comes at an unaffordably high price, I have urged that the conclusion of Meta-NMA is false. Concerning NMA, it turned out that its central part, the claim that novel predictive success of a theory justifies its correspondence truth, faces an important challenge from the history of science: it proved to be exceedingly hard to find a conception of success that is both generous enough to encompass those entities that the realist most likely wants to commit herself to and still excludes those entities that the realist does not want to commit herself to.

Finally, I have shown that the conclusion of the second IBE of NMA—scientific realism with regard to a given theory *T*—falls clearly short of the standards for scientific theories that Psillos himself erects in the course of his response to LSC: scientific realism by no means delivers any novel predictions. Together with justificatory monism, which implies that philosophical theories should be evaluated by the same justificatory standards used for scientific theories, this implies that, by Psillos' own lights, scientific realism should not be considered as true with regard to any scientific theory.

To conclude, I submit that Laudan-Style Counterexamples merely accentuate a problem that has its roots in the very core of the attempt to give a scientific argument for scientific realism. As shown above (Sect. 8.1), the central inference of NMA is from the novel predictive success of a given theory *T* to the adoption of scientific realism with regard to this theory. For empiricist justificatory monists, the basic problem with this inference is that it is in no recognizable way justified with recourse to empirical-scientific evidence. As Mizrahi (2012) has noted, scientific realism does not make any novel predictions that are not derivable from alternative explanations for the empirical success of scientific theories. Among these alternative explanations, Mizrahi selects what he calls weak surrealism, which is based on the claim that "[t]he observable world behaves as if our mature scientific theories are true" (Mizrahi 2012, p. 133). It seems that the history of modern philosophy furnishes a number of other alternative explanations that could be mentioned as well, among them Leibniz's monadic universe or Kant's transcendental idealism. More recent explanations for the empirical success of science include empiricism as proposed in van Fraassen (1980) and epistemic instrumentalism, as proposed by Stanford (2006). All of them make at least the same amount of predictions of novel phenomena as scientific realism, namely at least zero.

In fact, empirical evidence of any kind, let alone novel predictions, is simply irrelevant when it comes to choosing any of these explanations for successful theories. Whether one chooses to be a Kantian idealist, a van Fraassian empiricist, a Leibnizian monadologist, or a Psillonian realist cannot be decided with recourse to empirical findings. Empirical findings cannot decide which of these competing explanations for the empirical success of science should be adopted.

But isn't the debate between realists and these various kinds of non-realists still empirical insofar as the theories proposed by all sides have to correspond to the current knowledge about scientific inquiry? Certainly, the realist is in danger of being contradicted by future developments in science: if scientists decide to abandon one of the realist's favorite theories, say, general relativity, then this will tilt the race of arms mentioned above in favor of the non-realist: the realist will then have to find a criterion that excludes general relativity from the class of metaphysically true theories.[13]

This is certainly correct. However, it is important to note that this is a very weak sense of 'empirical success'. Rather than requiring successful empirical predictions, it boils down to the requirement that a given theory fits the facts known at that time. Psillos (1999, p. 105) is very critical of the force of such a requirement: "any theory (and, for that matter, any wild speculation) can be made to fit the facts—and hence to be successful". Psillos is completely right: according to this notion of empirically successful, van Fraassen's constructive empiricism, Stanford's instrumentalism and, indeed, any wild speculation that is tuned to these known facts would count as empirically successful. In short, to claim that scientific realism is empirically successful in this sense is not to say much at all.

This outcome suggests that the empiricist justificatory monist, who urges that philosophy of science use the same justification that is also employed in empirical science, cannot defend scientific realism.

[13]As a matter of physical fact, general relativity is already contradicted by non-local quantum phenomena. Informally, non-locality on the quantum level can occur when two particles in a superposition are separated from each other. Regardless of the distance between the two particles, a measurement on one of them has instantaneous effects on the other particle. General relativity does not allow such effects to happen instantaneously: no cause can have a distal effect that takes place faster than it would take light to travel between cause and effect. For a lucid introduction, compare Albert (1992, pp. 61–72).

References

Albert, David Z. 1992. *Quantum Mechanics and Experience*. Cambridge: Harvard University Press.

Busch, Jacob. 2008. No New Miracles, Same Old Tricks. *Theoria* 74: 102–114.

Carroll, Lewis. 1895. What the Tortoise Said to Achilles. *Mind, New Series* 4 (14): 278–280.

Devitt, Michael. 1997. *Realism and Truth*. Princeton: Princeton University Press.

Fine, Arthur. 1991. Piecemeal Realism. *Philosophical Studies* 61 (1): 79–96.

Fitzpatrick, Simon. 2013. Doing Away with the No Miracles Argument. In *Perspectives and Foundational Problems in Philosophy of Science*, ed. V. Karakostas and D. Dieks, 141–151. Cham: Springer.

Harman, Gilbert H. 1965. The Inference to the Best Explanation. *The Philosophical Review* 74 (1): 88–95.

Iranzo, Valeriano. 2008. Reliabilism and the Abductive Defence of Scientific Realism. *Journal for General Philosophy of Science/Zeitschrift für allgemeine Wissenschaftstheorie* 39 (1): 115–120.

Josephson, John. 1994. *Abductive Inference: Computation, Philosophy, Technology*. Cambridge: Cambridge University Press.

Kuhn, Thomas S. 1996 [1962]. *The Structure of Scientific Revolutions*. Chicago: University of Chicago Press.

Ladyman, James. 2011. Structural Realism Versus Standard Scientific Realism: The Case of Phlogiston and Dephlogisticated Air. *Synthese* 180 (2): 87–101.

———. 2014. Structural Realism. In *The Stanford Encyclopedia of Philosophy*, ed. Edward N. Zalta. Stanford University. http://plato.stanford.edu/archives/spr2014/entries/structural-realism/. Visited on 08 June 2014.

Laudan, Larry. 1981. A Confutation of Convergent Realism. *Philosophy of Science* 48: 19–49.

Lipton, Peter. 2004. *Inference to the Best Explanation*, 2nd ed. London: Routledge.

Lyons, Timothy D. 2002. Scientific Realism and the Pessimistic Meta-Modus Tollens. In *Recent Themes in the Philosophy of Science*, 63–90. Dordrecht: Kluwer Academic.

———. 2009. Non-competitor Conditions in the Scientific Realism Debate. *International Studies in the Philosophy of Science* 23 (1): 65–84.

Mizrahi, Moti. 2012. Why the Ultimate Argument for Scientific Realism Ultimately Fails. *Studies in History and Philosophy of Science* 43: 132–138.

Newman, Mark. 2010. The No-Miracles Argument, Reliabilism, and a Methodological Version of the Generality Problem. *Synthese* 177 (1): 111–138.

Papineau, David. 1993. *Philosophical Naturalism*. Oxford: Blackwell.

Psillos, Stathis. 1999. *Scientific Realism. How Science Tracks Truth*. London: Routledge.

———. 2007. The Fine Structure of Inference to the Best Explanation. *Philosophy and Phenomenological Research* 74 (2): 441–448.

———. 2009. *Knowing the Structure of Nature*. Basingstoke: Palgrave Macmillan.

———. 2011. Choosing the Realist Framework. *Synthese* 180 (2): 301–316.

Putnam, Hilary. 1975. What Is Mathematical Truth? In *Philosophical Papers, Volume 1*, 60–78. Cambridge: Cambridge University Press.

Quine, Willard Van Orman. 1974. *The Roots of Reference*. LaSalle: Open Court Publishing Co.

Sankey, Howard. 2008. *Scientific Realism and the Rationality of Science*. Aldershot: Ashgate.

Stanford, P. Kyle. 2006. *Exceeding Our Grasp: Science, History, and the Problem of Unconceived Alternatives*. Oxford: Oxford University Press.

van Fraassen, Bas. 1980. *The Scientific Image*. New York: Oxford University Press.

———. 2002. *The Empirical Stance*. New Haven: Yale University Press.

Weisberg, Michael, Paul Needham, and Robin Hendry. 2011. Philosophy of Chemistry. In *The Stanford Encyclopedia of Philosophy*, ed. Edward N. Zalta. Stanford University. https://plato.stanford.edu/archives/win2016/entries/chemistry/. Visited on 5 August 2017.

9

Scientifically Defending Realism II: Maddy's Piecemeal Realism

In the preceding chapter, I have argued that the basic problem with holistic arguments for scientific realism such as Psillos' is that they require a non-empirical-scientific way of justifying claims and theories and are hence unacceptable to the empiricist justificatory monist. I now consider an attempt to defend a limited, piecemeal realism, which promises to stay much closer to actual scientific inquiry, and hence to be acceptable for the empiricist justificatory monist. The attempt is due to Penelope Maddy.

I first discuss the central historical situation that Maddy analyzes to support her position, namely a series of experiments conducted by Jean Perrin based on theoretical work by Albert Einstein at the beginning of the twentieth century. Based on this discussion, I analyze her account how realism regarding atoms is established by these experiments (Sect. 9.1). Then, I argue that realism about atoms is not established in an empirical way by these experiments (Sect. 9.2). Finally, I sum up my critique of this attempt to defend an immanent, piecemeal realism, and I relate the results of this present chapter to my findings regarding no-miracle arguments to argue that empiricist justificatory monists are indeed unable to defend scientific realism (Sect. 9.3).

R. Gubelmann, *A Science-Based Critique of Epistemological Naturalism in Quine's Tradition*, https://doi.org/10.1007/978-3-030-24524-5_9

Before discussing the central experiments, let me introduce Maddy's piecemeal realism in its broad outlines. Maddy details her basic worry with regard to holistic approaches like Psillos' in the following passage:

> Such a realist may hold that we come to accept a theory because it is the best explanation of phenomena, or because there would otherwise be no explanation of its success, or whatever – these ideas come up again in IV.5 – but the salient point for now is that the discussion is taking place at van Fraassen's epistemic level. What's at issue isn't the reliability of the Einstein/Perrin evidence, but the puzzling interpretive question van Fraassen poses, a question that floats free of the ordinary methods of science. Such a realist is not content to agree with the Second Philosopher that the Einstein/Perrin evidence establishes the existence of atoms; he wants to convince van Fraassen, who sets ordinary evidence aside, that atoms really exist! The Second Philosopher feels no need for this extra stamp of the foot. (Maddy 2007, p. 310)

In this passage, Maddy first introduces a realist who has a holistic conception of theory confirmation: he comes to accept a theory as correspondence true because it provides the best explanation of its phenomena, or because there is no other explanation for its success. Then, she points out what is most important from her point of view, which is the perspective of her idealized naturalistic philosopher-scientist, the Second Philosopher. What is central to the Second Philosopher is that this debate between realists and van Fraassen (who endorses a sophisticated version of empiricism) is epistemically strictly distinct from the debate that actually took place among scientists at the turn of the twentieth century: the evidence provided there, what Maddy refers to by the term "Einstein/Perrin evidence", is of no relevance to that debate between realists and van Fraassen.

At first sight, this agrees nicely with the result of my analysis of holistic arguments for scientific realism: there is an epistemic gap between the kind of evidence that scientists can muster and the kind of claim that scientific realism makes, i.e. that in addition to being beautifully supported by the available evidence, the theory is also correspondence true.

However, the passage also shows that Maddy's diagnosis differs from mine: she does not think that realism as such is responsible for the failure of the holistic arguments as scientific arguments for scientific realism. Rather, she thinks that it is the manner in which the holistic arguments proceed; they agree with non-realists like van Fraassen to debate the question on a level where what she calls ordinary evidence has no bearing. According to her, it is precisely this move that seals the fate of any such naturalistic attempt to defend realism: "Once this move is made, the game is lost, because the only compelling evidence has been officially set aside as ineffective for these higher purposes" (Maddy 2007, p. 311). Hence, while I have suggested that it is the doctrine of realism itself that is beyond the justificatory monist's reach, Maddy here holds that the problem lies in the sort of non-realist that the arguments are trying to convince: non-realists like van Fraassen who disregard the ordinary scientific evidence, Maddy urges, cannot be reached by a truly scientific argument that restricts itself to ordinary scientific evidence.

This limitation notwithstanding, Maddy maintains that it is possible to scientifically defend scientific realism in a piecemeal fashion:

> Second Metaphysics is emphatically not a purely descriptive enterprise: the Second Philosopher holds that the existence of atoms had not been established in 1900 and that it had been established by 1910, not merely that scientists thought this or that at various times. Where it differs from the old normative project is not in renouncing normativity, but in its piecemeal approach: it doesn't begin with the demand for a general criterion. So, how *does* the Second Metaphysician proceed? For her, the answer to 'what is there?' takes the form of a list; what she actually confronts are a series of particular existence questions. (Maddy 2007, p. 403)

In this passage, Maddy first notes that her naturalized version of metaphysics, what she calls 'Second Metaphysics', consists in more than mere descriptive sociology (which would be restricted to record the opinions of scientists at some time and place). Rather, the Second Metaphysician can make fully-fledged ontological claims, and she can defend them such that rational readers should accept them. Furthermore, Maddy suggests that justificatory monists like Psillos make a tactical mistake: they try to defend their realism on too general a level. Instead, Maddy urges a piecemeal procedure: make a list of the things of which you would like to know whether they are real or merely useful fictions, and then assess them on a case-by-case basis. Her most important case in point is the question of the existence of the atom, which she thinks was scientifically settled by Perrin at the beginning of the twentieth century (with theoretical support from Einstein).[1]

9.1 Perrin's Experimental Case for the Reality of the Atom: Maddy's Case in Point

In this section, I focus on a famous series of experiments conducted by the French scientist Jean Perrin, who built on theoretical work by Albert Einstein. These experiments convinced a number of influential critics of atomism, such as Henri Poincaré or Wilhelm Ostwald[2] of the reality of the atom. Maddy focuses on these experiments to exemplify her piecemeal realism. Hence an analysis of her reading of the experiments as providing direct evidence for the reality of the atom will show whether her piecemeal realism succeeds where Psillos' holistic proposal has failed.

In the following, I detail the experiments that Jean Perrin conducted to verify predictions by Albert Einstein (Sect. 9.1.1). Then, I reconstruct Maddy's case for her claim that Perrin thereby provided a purely scientific case for scientific realism regarding atoms (Sect. 9.1.2).

9.1.1 What Is the Einstein/Perrin Evidence?

In the passages quoted at the beginning of this chapter, Maddy repeatedly alludes to *the Einstein/Perrin evidence* for the existence of atoms. In

[1]Wolff (2015) presents an anthropological challenge to the piecemeal approach of the Second Philosopher: she argues that, being the limited beings that we are, we cannot examine each question and position in detail and hence need general rules of Psillos' kind to orient ourselves. This is certainly an interesting point. However, the Second Philosopher could respond to this objection simply by hiring the required number of scholars to work down the list on a case-by-case basis.

[2]Poincaré's views on the issue will be detailed below, Sect. 9.2.2. The main aspect of Perrin's experiments that convinced Ostwald was that they simultaneously corroborate the molecular kinematic theory and clearly contradict its main rival, thermodynamics. See Ostwald (1912, p. 485).

this section, I discuss this evidence, following the lead of Mary Jo Nye's *Molecular Reality* (Nye 1972).[3]

According to Chalmers (2011, p. 724), and also, even if less explicitly, according to Nye (1972, pp. 102–136), Perrin's experimental case for the reality of molecules (and hence of atoms) rests on three main pillars. First, an experiment confirming the exponential vertical distribution of granules suspended in a fluid, yielding (as an aside, as it were) another, very precise way of computing Avogadro's number (Psillos 2011, pp. 351–358).[4] Only after obtaining these results did Perrin seriously engage the theoretical work of Einstein. Thus, he realized that the results from this first experiment confirm Einstein's prediction (Nye 1972, p. 111). He then proceeded to verify two further predictions of Einstein's, namely the random displacement of particles and their rotation energy (Nye 1972, pp. 126–136).

In the following, I focus on the first experiment and very briefly address the second one. The purpose of my detailed discussion of this first experiment is to give a firsthand impression of the sort of evidence for atomism that Perrin could provide.

Exponential Height Distribution. This first batch of experiments conducted by Perrin, and published in Perrin (1909) concerns the height distribution of granules suspended in a fluid of lesser density. The entire theoretical framework in which he conducted the experiments was innovative, to say the least, given the state of knowledge of the time. His entire set-up, and the painstaking, time-consuming experimental work, depends on an analogy between the behavior of suspended particles in a liquid and gas particles (Nye 1972, pp. 103–104). The notion of osmotic pressure was well-known at the time: given a container divided up in two compartments by a semi-permeable membrane that is permeable to the particles of the solvent, say, water, but not permeable to the substance dissolved,

[3]Being a committed advocate of Perrin's position, Nye by no means maintains a neutral perspective on the debate and, as will be shown below, sometimes distorts the views of Jean Perrin's peers. However, when it comes to retelling the story of Perrin's own experiments, which largely confirmed predictions by Albert Einstein, her account is still canonical for its attention to detail and careful analysis of the research context of the time. Among many others, her research is related by Gardner (1979), Egg (2014, p. 95), and by Maddy (2007, pp. 404–407).

[4]Avogadro's number specifies the number of molecules per mole, it is approximately 6.022×10^{23} per mole.

say salt, and given that the concentration of salt is significantly higher in one of the two compartments, the salt particles will exert an osmotic pressure against the membrane because the system aims at a state where the distribution of salt is even.

The basic idea behind Perrin's height distribution experiment is that the same holds for the vertical distribution of granules in a liquid: all things being equal, these granules will distribute evenly among the entire volume. However, if the granules have a greater mass density than the surrounding liquid, gravity will pull them down to the bottom of the cylinder. Against this gravitational pull, Perrin predicted that there was an osmotic pressure working to achieve an even distribution of the granules across the volume. As a consequence, the granules would not simply all drop down to the ground sooner or later, but rather show a systematic height distribution: with increasing height, the concentration of granules in the liquid decreases.

As Einstein (1905, p. 550) emphasizes, this directly contradicts the prediction by thermodynamics, as understood at the time: according to it, no osmotic pressure would operate on the granules, they would simply fall to the ground. The reason for this difference in prediction is simple (see Nye 1972, p. 98): if there is an osmotic pressure operating on the granules (and preventing them from dropping to the ground), this means that unobservable, but causally efficacious, particles must, as it were, continuously bump into the granules. Hence, verifying Perrin's predictions would at the same time provide positive evidence for the molecular kinetic view on matter and negative evidence against its main competitor, thermodynamics. Note that evidence is here understood in a holistic, Quinean sense: the system behaves as predicted by atomism, but not by thermodynamics.

Chalmers (2011, p. 720) puts the relevant correlation that Perrin derived as follows:

$$W \times \log\left(\frac{n_0}{n}\right) = 2\pi r^3 \Delta g h \tag{9.1}$$

where

- W the mean kinetic energy of a particle (units $\frac{kgm^2}{s^2}$)

- n_0 the density of particles at a given height x (units $\frac{kg}{m^3}$)
- n the density of particles at height $x + h$ (units $\frac{kg}{m^3}$)
- r the radius of the particles (unit m)
- g the gravitational acceleration (units $\frac{m}{s^2}$)
- h the distance between the two concentration-measurements n_0 and n (unit m)
- Δ the positive density difference between the particles and the fluid (units $\frac{kg}{m^3}$).

Intuitively, this equation describes a state where a particle's osmotic pressure is in balance with its gravitational pull. On the left-hand side of the equation is the osmotic pressure. $\frac{n_0}{n}$ is the density gradient, that is, the extent of the density difference between two points in the cylinder separated by height h. Because the density function $\log\left(\frac{n_0}{n}\right)$ is logarithmic, density decreases exponentially with increasing height. The stronger the differences in density, the higher the gradient and the higher the osmotic pressure exerted on the granules. Similarly, the higher W, the mean kinetic energy per particle, the more the particle are going to bump around and hence work towards an even distribution. On the right-hand side of the equation is the effective pull exerted on a particle by gravitation. More precisely, it's 3/2 of the volume of the sphere, multiplied by the positive density difference between the granules and the surrounding liquid and the gravitational acceleration, again multiplied by the height difference between which the density difference was measured.

Perrin was not only able to verify that the density decreases exponentially with increasing height as predicted by this equation, he was also able to measure Δ and r and thus had all the information to calculate W, the mean kinetic energy per unit volume of particles at a given temperature (Chalmers 2011, pp. 720–721).

It is worth pausing to appreciate the experimental genius and sheer perseverance Perrin had to muster to conduct these experiments (compare Nye 1972, pp. 105–109). The first challenge was simply to obtain suitable particles, or granules—they had to be very small (less than one micrometer in diameter, one micrometer is 0.001 mm), of the same size, and perfectly spherical. He prepared them using gamboge, a dried vegetable latex. First, he rubbed a stick of gamboge by hand into water, which created a yellow emulsion made up of glass-like, spherical grains. Then, he applied methyl alcohol that would dissolve 4/5 of the weight of the material. Finally, the central step consisted in complicated and iterated centrifugations to obtain granules of the same exact size.

The greatest challenge, however, was simply to count the particles to determine densities. For the first experiments, Perrin used an ultramicroscope, after that, he switched to ordinary microscopes. Perrin's method for counting the granules at a given height involved a needle-point field that divides up the visible field into very small units, where one could count the number of granules in an instant. Repeating this about 200 times per measurement yielded the required numbers of granules.

Up to this point, I have sketched how Perrin verified the exponential distribution of granules. Now, I focus on how he experimentally verified Avogadro's number. In a sense, he simply assumed that kinetic theory applies as much to his granules as it applies to gas molecules, as I intend to show now, following the lead of Chalmers (2011, p. 724). According to Chalmers (ibid.), Perrin, by simply presupposing that the granules would behave qualitatively and, for present purposes, *quantitatively* like gas molecules, was able to derive Avogadro's number from the mean kinetic energy. To do so, he used the following quantitative relationships that were well-known to hold for gases:

$$P = \frac{2}{3} \times \frac{n}{V} \times W \tag{9.2}$$

This equation represents pressure P (measured in *Pascal*, units $\frac{kg}{ms^2}$) exerted by a volume of gases, as a function the number of gas molecules n per volume V (measured in $\frac{kg}{m^3}$) and of the molecules' mean kinetic energy W (measured in *Joule*, units $\frac{kgm^2}{s^2}$): the more energy the molecules have, and the tighter they are packed, the harder they bump at a wall enclosing them.

Furthermore, Perrin presupposed the gas law:

$$PV = \frac{n}{N} \times RT \tag{9.3}$$

Here, P is again the pressure exerted and V is the volume of the respective gas portion. R is the gas constant (8.3144598 $\frac{\frac{kgm^2}{s^2}}{Kelvin}$ or, simpler, $\frac{Joule}{Kelvin}$) and T is the temperature (measured in *Kelvin*). Dividing this formula by V, the volume of the respective gas portion, and then equating it to the one immediately above (9.2), one gets, with small modifications:

$$N = \frac{3RT}{2W} \tag{9.4}$$

This Eq. (9.4) allowed Perrin to compute Avogadro's number from the gas constant, the temperature of the fluid and the mean kinetic energy that he could derive using the Eq. (9.1). According to Perrin (1909, pp. 59–61), the values he received for N were strikingly close to theoretical expectations for Avogadro's number. Even more, the value for N remained also constant across substantial variations of the observable parameters. Together with the fact of the exponential height distribution, this yields two elements of this batch of experiments that were generally seen as strongly supporting the molecular kinematic theory and, in some sense of 'real', the reality of the atom.

Displacement. I will only briefly address Perrin's visualization of the random walk of granules, to the extent to which it will become relevant later on. Perrin prepared the set-up with his usual experimental genius, producing granules from mastic (a kind of gum). Then, he recorded the position of each of three granules of mastic in 30-second intervals. This yields plots showing the path of three of these granules over time, in discrete units of 30 second intervals. The plots show intuitively that the displacements are random and independent of each other; more importantly, they neatly confirmed Einstein's prediction regarding the displacement of such granules. For more details, compare Nye (1972, pp. 132–135).

In sum, Perrin was able to verify Einstein's predictions regarding height distribution and displacement. Furthermore, he was able to experimentally derive Avogadro's number with very high precision and constant results across different experimental set-ups. In addition, the plots of the random movement of the granules provided immediate, intuitive support for Brownian motion.[5]

9.1.2 Maddy's Conception of the Significance of Perrin's Results

In this section, I discuss Maddy's reading of Perrin's experimental results, which I have presented in the preceding section. I first introduce her claim that Perrin's results constitute a new kind of evidence for the existence of atoms, what she calls 'thick evidence'. Then I detail her refusal to put any semantic or epistemic weight on her notion of thick evidence, and her insistence that it is the experimental results themselves that establish realism regarding atoms.

Maddy (2007, pp. 398–400) distinguishes as follows between *thin* and *thick evidence.* Thin evidence corroborates entire scientific theories, and it is broadly analogous to Quinean holistic theoretical virtues: successful empirical prediction, simplicity, etc. Thick evidence, however, supports particular existence claims. One species of thick evidence is direct observation. If zoologists in Siberia spot a mammoth (and perhaps take some pictures of it), this is direct evidence for the existence of mammoths in our day. Another species of thick evidence is what Maddy tentatively calls detection. It is tailored to the phenomenon of so-called theoretical or unobservable entities.[6] Maddy explicitly conceives Perrin's achievement as amounting to detection of atoms:

[5]He also verified an Einsteinian prediction of the mean rotation energy of particles, which I have not discussed here, compare Nye (1972, pp. 129–130).

[6]It is of course highly contested what phenomena count as observing: for instance, do microscopes count? However, this is not a problem that Maddy faces, as she explicitly does not put any epistemic weight on some event's counting as a detection or observation, see below in this section. For an early critique of the distinction, see Maxwell (1962). In an earlier work, Maddy does use 'seeing', albeit in scare quotes, in her description of Perrin's achievements: "atomic theory around the turn of the twentieth century was well-endowed with the five theoretical virtues, but it was not accepted as true until it had passed a further test. That further test was sometimes described in terms of 'seeing' or 'observing', other times in vaguer terms of 'direct testing' or 'experimentally verifying'. I won't attempt to analyze this notion; [...]" (Maddy 1997, p. 142).

> Now I've been suggesting all along, on the Second Philosopher's behalf, that the evidence involved in establishing the atomic hypothesis wasn't just more of the same, but a new type of evidence altogether, what we've been calling detection. (Maddy 2007, p. 405)

On the following page, she gives her most detailed account of why Perrin's experiments constitute a new type of evidence, one that, as Maddy (2007, p. 404) emphasizes, the previous accounts of the debate—Quine's, Boyd's, Cartwright's, Achinstein's, etc.—failed to capture (quoting her quoting Perrin is necessary because otherwise the full force of her point is lost):

> Before Perrin's successes, the case for the existence of atoms had hinged on aggregate behavior: the hypothesized activities of large groups of molecules were used to explain phenomena from combining volumes of heat. [...]
>
>> Molecular movement has not been made visible. The Brownian movement is a faithful reflection of it, or, better, it is molecular movement in itself, in the same sense that infra-red light is still light. From the point of view of agitation, there is no distinction between nitrogen molecules and the visible ... (sic) grains of an emulsion. (Perrin 1913, p. 105)
>
> Thus we are privy not only to a striking aspect of individual behavior, but in fact to its most counterintuitive aspect – the random walk – the very source of the conflict with classical thermodynamics. I think these considerations give some sense of how Perrin's evidence was qualitatively different from what came before, and why it was rightly regarded as a watershed. (Maddy 2007, p. 406)

Maddy claims here that (1) we are able to witness individual behavior. Given the context, she must be referring to the individual behavior of granules ingeniously prepared by Perrin in an emulsion. Furthermore, she claims that (2) we are able to witness the random walk of these granules, which conflicts with classical thermodynamics. There are two candidates for the 'qualitative difference' of this evidence. The first one is the fact that we are able to track individual granules. The second one is the fact that Perrin's results, by recording the random walk of granules, clearly created problems for thermodynamics, the main competitor of the molecular theory at the time.

Maddy never clearly states what exactly qualifies Perrin's experimental results as detections of atoms (or molecules), and hence as thick and not only thin evidence. This is not a consequence of sloppy work; it rather represents her well-considered position. Maddy emphasizes that she does not have a clear-cut conception of detection, and that the criterion is epistemically weightless: she calls her notion of detection "vague and/or easily disposable" (Maddy 2007, p. 402). She argues that giving a clear-cut criterion with epistemic weight would not only be contradicted by scientists' amazing creativity in devising ever new ways of detecting things, it would also tempt philosophers of science to put the epistemic cart before the horse:

> In other words, the ordinary science, not the criterion, is what's doing the work; [...] This isn't to say that the scientist's judgment can't be questioned – all her evidence and reasonings are open to perfectly ordinary scrutiny and critique – but the criterion alone isn't providing any leverage. So I wonder: perhaps the sort of toothless criterion under consideration here – with its vague and/or easily disposable notion of 'detection' – isn't a fully-fledged discourse criterion in Azzouni's eyes. (Maddy 2007, p. 402)

According to this passage, the fact that Perrin's experimental achievements qualify as detection of atoms does not establish anything with regard to the existence of atoms, it does not justify accepting them as real; rather, it is the experimental achievements as such that corroborate the reality of atoms.

This refusal of Maddy both to clarify her conception of detection and to put any epistemic weight on it is central to her piecemeal realism. What happens if you have a clear-cut and epistemically weighty criterion to decide whether some scientific posit should be taken to exist in reality has been studied on the example of Psillos' NMA: any scientific theory to which his conception of empirical success applies is thereby a theory to whose correspondence truth Psillos commits himself. Only because Psillos puts so much epistemic weight on his criterion of success does the conceptual race of arms get off the ground: anti-realists can then turn to the history of science and identify a theory that is both successful in Psillos' sense but also fundamentally mistaken according to contemporary theories.

Maddy's insistence on the exclusive relevance of ordinary scientific evidence can be seen as an attempt to block such a conceptual race of arms. As I have shown in the preceding chapter, in this race, particular aspects of Perrin's experiments and Einstein's theoretical work hardly come up when it comes to deciding whether we are justified to be realists regarding atoms. Rather, what is at issue in this race, to put it somewhat bluntly, is whether there is an early modern alchemist theory whose ontological commitments we find outrageous, but which does fulfill the latest proposed criterion of empirical success.

This implies that, for Maddy, the experiments sketched in the preceding section directly establish the real existence of atoms; her gestures at a "striking aspect of individual behavior" that we can witness thanks to Perrin's experiments, or the fact that the random walk of the granules contradicts thermodynamics, are only a means to allow us to appreciate the full force of the experiments, they do not themselves have any justificatory force. In the next section, I challenge this claim that the experiments themselves, from the background of Einstein's theoretical work, justify realism regarding atoms.

9.2 Critical Appreciation of Maddy's Piecemeal Realism

In this section, I challenge the central claim of Maddy's piecemeal realism: her claim that Perrin's experiments, as detailed above (Sect. 9.1.1), directly establish realism regarding atoms. If warranted, this claim would nullify the conclusion of the preceding chapter, according to which scientific realism is a doctrine that cannot be defended by empirical-scientific means. By the same token, it would show that empiricist justificatory monists can defend scientific realism, which in turn constitutes a decisive step towards coming to terms with the problem of justificatory pluralism (see above, Sect. 7.1).

My challenge of Maddy's central claim proceeds in three steps. First, I contrast Maddy's reading of the experiments with Bas van Fraassen's alternative reading of the significance of Perrin's experiments to bring to light the fact that Maddy needs a criterion with justificatory force to maintain realism regarding atoms (Sect. 9.2.1). Then, in Sect. 9.2.2, I reinforce this finding by showing that both Henri Poincaré and Jean Perrin did not adopt realism regarding atoms as a direct consequence of Perrin's experiments, but rather based on intricate arguments. I conclude by discussing alternatives both to my reading of Maddy's position (Sect. 9.2.3) as well as of the atoms debate (Sect. 9.2.4).

9.2.1 Why Maddy Needs a Criterion with Justificatory Force

The research presented by van Fraassen (2009) provides a detailed alternative to Maddy's reading of Perrin's results. Using Maddy's vocabulary, one can say that van Fraassen insists that Perrin provided (and even thought to provide) only thin evidence for the molecular kinematic theory, as opposed to thick evidence for the existence of the atom. To defend this claim, van Fraassen (2009, p. 21) develops the concepts of concordance and determinability, which together make up empirical grounding, the primary epistemic virtue that a scientific theory should possess. Concordance occurs when different ways of measuring a certain quantity agree with each other (van Fraassen 2009, p. 16), determinability is the requirement that it be possible to obtain a definite measurement for each significant parameter of a theory in each specific situation (van Fraassen 2009, p. 11). Then, van Fraassen argues that the work of Perrin achieved empirical grounding of the molecular kinematic theory for the first time and thereby justified this theory in the eyes of the scientific community.

What counts as detection, as establishing the reality of the atom, in Maddy's reading—the computation of Avogadro's number by different means, the verification of the vertical height distribution of granules, and the verification of the random walk of granules—amounts to empirical grounding in van Fraassen's interpretation. The ten-dollar question now is of course: who is right? Whose interpretation of Perrin's results should be adopted? More to the point, we can picture van Fraassen as asking Maddy for a justification of her claim that Perrin's experiments establish the reality of the atom. Given Maddy's insistence that she is not out to record "merely that scientists thought this or that at various times" (compare above, the introduction to the present chapter, Maddy 2007, p. 403), she cannot be content with pointing out that many of Perrin's fellow physicists took it as such. Rather, she has to give features of Perrin's results that make it rationally compelling to take them to support the reality of the atom.

Unfortunately, as has been delineated above, Maddy does not want to give such an answer. To give one would mean that the criterion itself is *not* vague and easily disposable, that it does provide epistemic leverage. To see why, assume that Maddy would respond to van Fraassen's request for a justification of her realist reading of the significance of Perrin's experiments: "Perrin's results amount to detection of atoms because Perrin was able to trace the individual behavior of granules." To this, the natural rejoinder would be: "Why is the tracing of individual granules sufficient to corroborate realism about atoms?" One can easily see that these questions, and others like it, force Maddy to provide reasons why detections corroborate existence claims, which means that she will have to attribute epistemic leverage to the criterion and give reasons why this criterion should have such epistemic leverage. This would be the beginning of a conceptual race of arms entirely analogous to the one in which Psillos got caught up. Perrin's experiments could not contribute much to this race of arms.

This means that Maddy must either give up *any* talk of thick evidence or of Perrin's experiments constituting a new kind of evidence, and hence let go of her realist claims, or give this talk the substance she thinks she cannot give to it.[7]

[7] *Prima facie*, there may seem to be a third option for Maddy here. It consists in the claim that Perrin's experiments made atoms observable *sensu stricto*. However, this would amount to reconceiving the question of the reality of the atom as a question for common-sense realism. Thus, this move would bypass rather than address the central issue of scientific realism, namely whether we can assert the reality of so-called theoretical, or unobservable entities. Furthermore, it does not seem convincing to conceive the achievement of Perrin's experiments as making atoms observable in the strict sense. As Egg (2014, p. 95) points out, "a Brownian particle undergoes some 10^{21} collisions with molecules of the surrounding fluid per second." Given the technology available by the time of Perrin's experiments, this is just too fast to observe in any substantial sense of "observe". None of this is to dispute that atoms have become observables now. I am only claiming that Perrin did not make atoms observable.

Maddy may feel that she can happily embrace the first option, while still being a realist about atoms: She could simply point to Perrin's experiments as single-handedly establishing the reality of the atoms, and just hope that van Fraassen will finally come to reason and recognize the obvious. After all, she has affirmed that it is the ordinary science that is doing the justificatory work needed to corroborate the existence of atoms.

However, and this seems to be the core of the matter, it is simply wrong that the evidence provided by Perrin's experiments directly establishes the existence of atoms and molecules in reality. Let us reconsider the two experiments on which Maddy capitalizes, as sketched above (Sect. 9.1.1). First, Perrin confirmed that the vertical height distribution of granules in a solution of lesser density is indeed exponential. At the time, the only available explanation for the granules' not simply falling to the ground was that there is an osmotic pressure operating on them. The only available explanation, in turn, for this osmotic pressure were unobservable molecules, as it were, bumping into the granules. By means of the same experiment, he was able to compute Avogadro's number N consistently in a number of different experimental circumstances.

The second batch of experiments established that the granules do indeed move around according to no systematic pattern, that is, randomly, thus creating serious problems for the thermodynamic account of matter, while neatly confirming a prediction made by Einstein based on the molecular kinematic approach. It seems clear that without further argument, none of these experiments directly implies the claim that atoms exist in mind—and theory-independent reality.

Rather than directly establishing realism regarding atoms, these results simply drastically increase the empirical support of molecular kinematic theory. Whether, in addition to that, they establish the existence in reality of atoms or molecules is a different question, a question that concerns the metaphysical status that we attribute to important posits of empirically highly successful theories; more particularly, in the case of vertical height distribution, the question is what metaphysical dignity we assign to theories justified by inference to the only conceivable explanation.

In light of this, what Maddy needs is a line of reasoning of the following form:

P1 Perrin's results establish that atoms, considered as posits of the molecular kinematic theory, have property *p*.
P2 Entities posited by science that possess property *p* exist in reality.
K Atoms exist in reality.

Within the framework of molecular kinematic theory, it has to be ascertained experimentally that atoms possess a certain property *p* (*P1*). This step can take place entirely within empirical science. Then, however, something of the sort of *P2* is needed—I call it a *bridge principle*. The purpose of the bridge principle is to bridge the justificatory gap between the empirical results and the metaphysical conclusion; hence, bridge principles are like Maddy's criteria—just that the bridge principles, unlike the criteria, do contribute substantially to the justification of realism about the respective scientific posit.

My justification for the need of a non-empirical bridge principle of the sort of *P2* is twofold. First, a close look at Perrin's experiments shows that they alone cannot warrant realism regarding atoms. Second, and building on this, the brief contrast with van Fraassen's reading of the significance of Perrin's experiments shows that a "vague and/or easily disposable" (Maddy 2007, p. 402) bridge principle will not do. If Maddy wants to defend her claim that Perrin's experiments justify realism about atoms against van Fraassen's alternative interpretation of these experiments, she needs to say what aspect of the experiments proves her interpretation right and van Fraassen's wrong. This cannot be done without endowing this aspect with the justificatory force to support her realism.

However, as soon as you have a statement of the general form of *P2*, you can start a conceptual race of arms analogous to the one provoked by Psillos' conception of empirical success: anti-realists can then scrutinize the history of science to find posits that did possess *p*, but that do not exist according to current theories. This debate then may still be empirical in a historical sense, but this is of course qualitatively different from empirically predicting and verifying hypotheses in physics. In regard to this, I entirely agree with Psillos' insistence that any wild speculation can be empirical insofar as it fits the facts, but not any theory can successfully predict novel phenomena (compare the discussion at the end of Sect. 8.3).

To conclude this section, I wish to emphasize that this does not mean that the question whether or not Perrin's experiments do establish the real existence of atoms is meaningless, or somehow misguided in principle. It just means that it is wrong to say that Perrin's experiments as such settle this question—which contradicts Maddy's reading of the experiments.

9.2.2 Perrin's and Poincaré's Bridge Principles

In this section, I show that two of the leading figures of the debate that was initiated by Perrin's experiments, Jean Perrin and Henri Poincaré, did not think that Perrin's experiments directly confirm the reality of atoms. Rather, both of them proposed (significantly different) arguments for the reality of the atom that rely on the experiments, but that clearly transcend anything that is directly implied by the experiments. This means that they proposed their versions of bridge principles to get from the experimental results to realism regarding atoms.

Perrin: The Atom-Evangelist. In the following, I consider key passages where Perrin reflects on the significance of the results of his famous experiments. Based on this, I argue that his criterion to decide whether or not a certain experiment justifies realism regarding a given scientific posit is concordance (in a slightly different understanding than van Fraassen's) together with a version of causal eliminative reasoning. In doing so, I am following the lead of Achinstein (2002).

Here is how Perrin conceives the results of his first batch of experiments concerning the confirmation of the exponential height distribution[8]:

> I do not think that this concordance can leave any doubt about the origin of the Brownian movement. To appreciate how significant this concordance is, one has to take into account that before this experiment, one would certainly not have ventured to affirm that the drop in concentration would not be negligible considering the miniscule height of some microns; similarly, one

[8] All translations from French to English are mine, unless otherwise stated. I am grateful to Christian Weibel for commenting on draft versions of the translations.

would never have ventured to claim that all the granules would not end up all together at the bottom of the container. According to the first hypothesis, the value of N' would be expected to be zero, according to the second it would be infinitely high. That, within this enormous interval that seems a priori possible for N', one ends up exactly with a value so close to the expected figure, does not quite appear to be a coincidence.

Hence, the molecular theory of Brownian movement can be considered experimentally established; by the same token, it becomes difficult to deny the objective reality of molecules. (Perrin 1909, p. 61)

Here we find the term 'concordance', but the context makes clear that Perrin uses it differently from van Fraassen. Immediately before the passage quoted, Perrin writes: "Mais, de plus, il est manifeste que ces valeurs concordent avec celle que nous avions prévue pour l'énergie moléculaire" (Perrin 1909, p. 61). Here, concordance is between experimental results and theoretical predictions. So, what Perrin calls concordance in the passage quoted is the remarkable agreement between the theoretical predictions and the experimental results.

More precisely, Perrin emphasizes that, apart from presupposing molecular kinematic theory, one could have expected any value between zero and infinitely high for *N*. This is because it is plausible to expect that the density gradient is at about zero, as the height difference he used in his measurements was only some microns. Furthermore, Perrin urges, the opposite assumption is also a priori plausible. It is plausible that this density gradient is very high, leading to a very high value for *N*, at the point where the density changes from 100 to 0%. That there is such a point is plausible because one would expect that the granules all end up at the very bottom of the cylinder, given that they are denser than the surrounding liquid. Thus, Perrin argues, the conceivable values for *N*, based on his experiments, could be anywhere between very, very high and very, very low. This makes it all the more striking that the actual value obtained was very close to Avogadro's number.

Furthermore, Perrin (1909, pp. 59–61) emphasizes that the value for *N* remained almost constant across experiments undertaken with strikingly different densities, masses, and rarefactions of granules. He also uses "concordent" in this context, which is similar to, but still different from van Fraassen's conception of concordance, since van Fraassen requires that different ways of experimentally asserting a quantity have to agree. In contrast, Perrin always used the same kind of experiment, just with different parameters.

Perrin takes these consistent results as strong evidence both for the molecular kinematic theory of Brownian movement and for the objective reality of molecules. This means that Perrin here advocates a holistic variety of scientific realism, similar to Psillos'. Why is it hard to deny the objective reality of the molecule? Because the theory in which they figure ineliminatively is tremendously successful in its empirical predictions: it has consistently, and in very different circumstances, received almost perfect confirmation.

Achinstein (2002, pp. 473–476) as well as Nye (1972, pp. 126–136) agree that the concordance, in Perrin's understanding, that Perrin observed in his experiments contributed to his conviction that his experiments establish realism regarding atoms. In addition, Achinstein (ibid.) claims that Perrin employs a second bridge principle, one that is very similar to contemporary causal eliminativist accounts. The basic idea of this kind of account is the following. We have an effect E and know that $C_1,\ldots,C_n$ can cause E. We then (perhaps experimentally) eliminate all of these known possible causes except C_1. According to the causal eliminativist, this justifies acceptance of C_1 as real.

To support his claim that eliminative causal reasoning is explicitly invoked by Perrin, Achinstein (2002, p. 490) quotes the following passage (translation mine):

> When we study a machine, we do not restrict ourselves to reason about the visible parts, which nevertheless only have reality to us, as long as we cannot disassemble the machine. We will do our best to observe the visible parts, it is true, but we will also seek to discover which mechanisms, which hidden organs explain the visible movements.
>
> To discover in this way the existence or the properties of objects which are still beyond our knowledge, and to explain the visible complexity with recourse to invisible simplicity, this is the kind of intuitive intelligence to which we owe atomism – thanks to men such as Dalton or Bolzmann – of which this book is an exposition. (Perrin 1913, pp. iv–v)

In this passage, Perrin draws an analogy between the use of molecular theory in science and our approach at finding out how a machine works. He maintains that we do not restrict ourselves to the machine's visible parts, but do make conjectures regarding the invisible ones that explain the behavior of the visible parts. Furthermore, Perrin specifies that we should use *simple* explanations with recourse to invisibles to explain the *complicated* behavior of the visibles, and he submits that this method has given us atomic theory. This passage provides strong evidence that his realism involves an explanatory element: we are justified to posit unobservable entities if they provide simple explanations of visible behavior. However, it does not as such provide support for the further exegetic claim that, for Perrin, this explanation must be *causal*. This support is provided by the following passage:

> On the contrary, it was established by the work of Mr. Gouy (1888) not only that the hypothesis of the agitation of molecules furnishes an admissible explanation of the Brownian movement, but also that one could not imagine another cause of this movement. This substantially increased the importance of the hypothesis. (Perrin 1909, p. 9)

Here, Perrin first states that the molecular kinematic theory explains Brownian motion satisfactorily, but also that it is the only imaginable cause of this movement. Thus, he easily moves from an admissible explanation to a causal one. Furthermore, the context is highly relevant: if it is true that the molecular kinematic explanation is the only admissible explanation for Brownian movement, and if Perrin succeeds in showing the reality of Brownian motion, then he has all the material in hand for a causal-eliminative inference to the reality of the molecule (and hence of the atom). Thus, I suggest that Perrin effectively uses this combination of concordance and causal eliminative reasoning.

In short, according to Perrin's own view, which is, in more sophisticated versions, still held by contemporary researchers in the field,[9] his experiments establish the reality of the atom through a combination of causal eliminative inference and the fact that his calculations for Avogadro's num-

[9]Compare Egg (2014).

ber agree to an amazing degree. The general argumentative form suggested above (Sect. 9.2.1) can therefore be specified in the following form. Its bridge principle *P2* is obviously non-empirical:

P1 Perrin's results establish that there is a phenomenon whose only conceivable causal explanation involves atoms of molecular kinematic theory.
P1a Perrin's results establish concordance (in Perrin's understanding) of Avogadro's Number *N*.
P2 Entities postulated by science that are the only conceivable causal explanation of a given phenomenon and for which concordance (in Perrin's understanding) is fulfilled exist in reality.
K Atoms exist in reality.

Poincaré: A Convert, Sort of Maddy (2007, p. 95) mentions Henri Poincaré as one who was "immediately convinced" by Perrin's experiments of the reality of the atom. Here is an excerpt from a that talk Poincaré held at the conclusion of the 1912 conference of the *Societé française de physique*, where Perrin's experimental results, as sketched above (Sect. 9.1.1) were at the center of attention:

> A first reflection must have struck all listeners; the former mechanistic and atomistic hypotheses have recently received sufficient consistency to almost cease to appear to us as hypotheses; atoms are no mere useful fiction anymore; it seems to us that we see them now, as it were, since we know how to count them. (Poincaré 1912, p. 347)
>
> The brilliant determinations of the number of atoms provided by Perrin have completed this triumph of atomism. What grounds our conviction are the multiple concordances between the results obtained by means of completely different procedures. Not long ago, one would have considered oneself lucky if the resulting numbers would contain the same number of digits; one would not have demanded that the first significant figure match; today, this first significant number has been established; and what is remarkable is that we have turned our attention to the most diverse properties of the atom. In the procedure that derives from Brownian movement, or in the one that invokes the law of radiation, it is never the atoms that are counted directly, [but rather] the degrees of freedom; in the experiments where one operates with the blue of the sky, the mechanic properties of the atoms do

> not matter anymore, instead they are rather conceived as causes of optical discontinuity; finally, if one makes use of Radium, what one counts is the emission of projectiles. In consequence, if there had been discordances, one would not have had difficulties to explain them, but fortunately there were none.
>
> The atom of the chemist is now a reality; but this is not to say that we are close to touching upon the ultimate elements of reality. (Poincaré 1912, pp. 348f.)

In the first passage, Poincaré expresses his conviction that the ancient mechanistic and atomistic hypotheses have almost ceased to be hypotheses, and that the atom is no more a practical fiction, as it seems, as it were, that we can see them now, since we are able to count them. In the second passage quoted, which is unfortunately ignored in the rather selective quote on Nye (1972, p. 157), he gives his real reasons for accepting the reality of the atom. While he uses a hedging term in the first passage—"pour ainsi dire" (as it were)—to qualify his claim that it appears that we see atoms now, there is no hedging in the second passage. There, he declares that it is the concordance of results obtained by completely different procedures and engaging very different properties of the atom that convinced him. He then elaborates on this point. Until recently, he reports, one would have been happy if the numbers—the number at issue is Avogadro's number—would have been on the same magnitude. One would never have demanded agreement on even the first significant figure. Poincaré then asserts that this has been achieved, and, most importantly, achieved by entirely different experimental means.

Let me give you the full quote from Nye, to evince why I am worried about her selection:

> A first reflection is sure to strike all listeners; the long-standing mechanistic and atomistic hypotheses have recently taken on enough consistency to cease almost appearing to us as hypotheses; atoms are no longer a useful fiction; things seem to us in favor of saying that we see them since we know how to count them. … The brilliant determinations of the number of atoms made by M. Perrin have completed this triumph of atomism. … In the procedures deriving from Brownian movement or in those where the law of radiation

> is invoked … in … the blue of the sky … when radium is dealt with. … The atom of the chemist is now a reality. (Nye 1972, p. 157)

First, she ignores the hedging phrase "pour ainsi dire" in her translation of the first paragraph quoted above. This is unfortunate, as Poincaré obviously wants to make clear that he does not put any weight on his use of the verb "to see". What is much worse however, is that she complements this ignoring of the hedging phrase with completely leaving out the part where Poincaré makes clear what convinced him: "Ce qui entraîne notre conviction, ce sont les multiples concordances entre des résultats obtenus par des procédés entièrement différents." Considering this sentence, it is clear that what convinced Poincaré to attribute reality to the chemist's atom is concordance in a sense that is even stronger than van Fraassen's: measurements based on the mechanistic properties of atoms (their degrees of freedom, as in the Brownian motion), others based on their optical properties, and even others on their property of emitting projectiles all agree on the number *N*. This convinced Poincaré of the reality of the atom. Hence, Poincaré's argument for the reality of atoms has the following structure. Again, the bridge principle *P2* is obviously non-empirical:

P1 Perrin's results establish concordance (in Poincaré's sense) regarding the atom of molecular kinematic theory.
P2 Entities postulated by science of which concordance (in Poincaré's understanding) is established exist in reality.
K Atoms exist in reality.

I wish to emphasize that Ivanova (2013) argues convincingly that this acknowledgment of the reality of the chemist's atom does not amount to a conversion of Poincaré to scientific realism, but rather to a reconsideration of the status of the atomic hypothesis in his Neo-Kantian philosophy of science. Poincaré changed the status of the hypothesis from an indifferent—that is, empirically insignificant—hypothesis to an empirical hypothesis, that is, an empirically testable hypothesis. This contradicts what Maddy (1997, p. 142; 2007, p. 95) suggests and what Psillos (1999, p. 22) and also Castro (2013) explicitly claim, namely that the passage quoted at length (Poincaré 1912, pp. 348f.) evinces Poincaré's conversion to scientific realism. Just one passage related by Ivanova (2015, p. 121) suffices to show how Neo-Kantian and how opposed to contemporary scientific realism Poincaré's thinking was:

> This harmony that human intelligence believes to discover in nature, does it exist outside of this intelligence? No, in all likelihood, a reality completely independent of the mind that conceives it, sees it or feels it, is impossible. A world so external, even if it did exist, would forever be inaccessible to us. (Poincaré 1905, p. 11)

In this passage, Poincaré makes it crystal clear that he thinks a mind- and theory-independent reality is inaccessible to human beings in principle. However, contemporary scientific realists argue that Perrin's experiments establish the existence of the atom in precisely such a reality.

Let me sum up this sketch of the reasons of Perrin and Poincaré to accept realism about atoms. Perrin himself thought his experiments do establish the reality of the atom because he believed that eliminative causal inference in conjunction with concordance in his understanding could justify realism about atoms. For Poincaré, concordance in his strong understanding is sufficient to establish realism about atoms—only that he did not mean by realism about atoms the mind- and theory-independent existence of atoms, but rather their becoming amenable to empirical investigation.[10]

9.2.3 Strict Epistemological Quietism: An Alternative Reading of Maddy's Epistemology

The gist of my critique of Maddy's piecemeal realism is that she impossibly tries to defend scientific realism regarding atoms without committing herself to the bridge principles (that are, as I have argued, both non-empirical and the starting point for a conceptual race of arms that will completely ignore the first-order scientific evidence provided by Perrin). Given her naturalism, this case is further complicated by the fact that eminent scientists of the time, such as Perrin and Poincaré, had no qualms about defending what I call non-empirical bridge principles.

[10] For Poincaré's views on ontology, see also Ladyman (2011).

While the passages examined so far in this chapter lead into this dead-end, I now briefly discuss a different understanding of her position, one that might escape this dead-end. In the following, I briefly sketch the two main parts of this alternative reading, or reconception, of her position, and I also point to its formidable price—a price, as I will urge, that Maddy cannot afford to pay.

The first part of this alternative reading centrally involves semantics, namely the semantics of 'detection'. Remember that I have urged that Maddy needs a clear-cut, substantial concept of detection, amounting to little less than a fully-fledged definition, to stand her ground against van Fraassen's objection that Perrin, rather than detecting atoms, merely provided empirical grounding for the molecular-kinematic theory as a whole.

There might be a way how Maddy can defend her reading of Perrin's experiments without initiating a conceptual race of arms that would inevitably leave the level of first-order scientific evidence. This way is prefigured in Tyler Burge's work on semantics. Consider the following passage, where Burge argues that, in addition to explicit, definitional ways to identify particulars that fall under a concept, there are more primitive, non-definitional ways to apply concepts:

> Concepts are not made what they are by understanding them. They are grounded in patterns of first-order usage. The usage has counterfactual implications. So it potentially extends to cases of application which never occur. Implicit in this potential is an ability to recognize the cases if they arise. With sufficient sophistication, commonly through language use, an individual may acquire the abilities to imagine counterfactual cases and to think up explications that are meant to cover all cases or simply to generalize about types of cases to which the concept would be applicable. These latter second-order abilities are epistemically posterior to the first-order abilities, even though they may come to accompany the first-order conceptual abilities implicitly. (Burge 2013, p. 518)

Before discussing this passage, remember that we are looking for a way to allow Maddy to insist on her claim that Perrin detected atoms, rather than merely providing empirical grounding for molecular-kinematic theory, without putting all of the epistemic weight on the conceptual specifications that she needs for this insistence.

According to the semantics sketched in this passage, this might be possible: she could conceive her identifying Perrin's experiments as detections as a first-order usage, and she could insist that any general explications of detection, while certainly helpful and necessary for the debate with van Fraassen, could not threaten her first-order usage, since the latter is epistemically prior to the former. This way, she could avoid the conceptual race of arms that any clear-cut concept of detection would initiate.

The central challenge with regard to this proposal is that the patterns of first-order usage regarding detection (understood as providing thick evidence) might not be substantial enough to furnish a sufficiently strong semantic (and ultimately epistemic) ground for Maddy to stand on and defend her usage of 'detection' vis-à-vis alternatives such as van Fraassen's. It might turn out that detection in this sense is a theoretical concept that, unlike 'tomato' (compare Burge 2013, p. 581), is not embedded in a thick pattern of usage and everyday acquaintance that could provide the ground for counterexamples and thought experiments with the potential to overturn explicitly held definitions.

Regardless of this more general worry, this semantics is not an option for the holistic approach chosen by Psillos (see above, Chapter 8). Psillos is not content to proceed on a case-by-case basis. Rather, he explicitly wants to confront anti-realists like van Fraassen on the general level of properties that, if possessed by any scientific theory, provide strong warrant for its correspondence truth. As a consequence, Psillos cannot do without putting all of the epistemic weight on these general criteria.

The second main part of this alternative reading consists in rejecting any substantial contrast between 'ordinary science' on the one hand and what Maddy calls 'criteria' and what I call bridge principles on the other hand. This would amount to accepting the atoms debate, at least the central aspects of it, as first-level scientific inquiry, as opposed to the second level of science-independent, philosophical theorizing that Maddy fundamentally rejects. A direct implication of including this debate is that Maddy cannot subscribe to empiricist justificatory monism anymore, not even of the implicit, unspecific kind that I have detailed above (Sect. 1.1.1). In epistemological terms, her position could be called a quietist one, not explicitly excluding empirical nor traditional metaphysical modes of inquiry. This position is at least implicitly suggested by a passage at the very beginning of her book, where she cautions against any clear-cut notions of science:

> [...] there is no hard and fast specification of what 'science' must be, no determinate criterion of the form 'x is science iff...'. It follows that there can be no straightforward definition of Second Philosophy along the lines 'trust only the methods of science'. Thus second Philosophy, as I understand it, isn't a set of beliefs, a set of propositions to be affirmed; it has no theory. (Maddy 2007, p. 1)

If, as this passage maintains, there is no clear-cut definition of science, then it might be difficult to maintain the clear distinction between ordinary science and criteria such as Poincaré's. This then leaves room for the quietist reading of her position.

As I see it, a first drawback of this alternative reading is that it threatens to extend the realm of the immanent, of the part of philosophy that is acceptable to the Second Philosopher because it is in touch with science, to a degree that makes Second Philosophy indistinguishable from First Philosophy. This is because the atoms debate is, epistemologically speaking, such a messy affair: empirical, metaphysical, epistemological, methodological, and even theological doctrines have influenced the participant's positions. And, at least prima facie, all of these doctrines would become fair game for the Second Philosopher if Maddy accepted this first suggestion.

In particular, the challenge would be to distinguish these doctrines and discussions from those that Maddy clearly wants to avoid, namely those that are conducted on a level that is, speaking epistemologically, entirely isolated from first-order scientific inquiry. In particular, Maddy (2007, p. 61) concludes that Kant's transcendental philosophy does amount to such a two-level conception that she strictly rejects: on the one level, the empirical one, there is ordinary scientific research, on the other, the transcendental one, metaphysical, epistemological and methodological theorizing occurs in a way that is entirely immunized against any insight or critique from the first level. This hermetic epistemic separation is what Maddy's Second Philosophy is fundamentally opposed to. Hence, the specific challenge will be to separate the inacceptable Kantian parts of Poincaré's position, those that are on the second, strictly non-scientific level, from those that are still in touch with the scientific level. It seems an open question whether this is possible.

A second and more serious drawback of this second part is as follows. Remember that an implicit commitment to a, perhaps rather unspecific, version of empiricist justificatory monism permeates her entire work. Abandoning this commitment would therefore require a reorientation of much of her research. For instance, Maddy would have to find another way to frame her rejection of Quine's developmental psychology, one not based on empirical evidence, and she could not subscribe to Mark Lange's prediction-based distinction between admissible and inadmissible metaphysics anymore (see above, Sect. 1.1.1).

In summary, this alternative reading of Maddy's position might allow her to avoid my critique: by conceiving of 'ordinary science' as involving the atoms debate in its entire complexity, or, perhaps more congenial to her position, by *not* adopting a conception that excludes this debate, she is entitled to partake in it and to accept its conclusion. Furthermore, by adopting a semantics that leaves room for epistemically weighty concept applications that are not guided by fully-fledged definitions, she might be entitled to conceive of Perrin's experiments as detections without putting all of the epistemic weight on definitions of the concept *detection*—and initiating a conceptual race of arms that she wants to avoid. What is clear is that Maddy's Second Philosophy, on this alternative reading, would then not be a version of empiricist justificatory monism anymore, regardless of how implicit and unspecific her adherence to this doctrine. Given the importance of this distinction for Maddy's work, I conclude that this alternative reading is not a real option for Maddy: the price to pay, abandoning even an implicit commitment to empiricist justificatory monism, is too high.

9.2.4 Was the Reality of Atoms Scientifically Justified Before Perrin?

The reading of the significance of Perrin's experiments developed here is contradicted in an interesting way by Castro (2013, pp. 44–53). Briefly, Castro holds that, by 1900, there was already purely scientific evidence that confirms the real existence of atoms. Hence, Perrin's experiments, and the debate provoked by them, was superfluous, the result of a failure of the scientists of the time to appreciate the fact that the question was already settled. In the following passage, Castro details the general thrust of his strategy:

> Overall, we should not focus on the scientist's opinions regarding their work, but only analyse scientific theories and experiments. For me, a naturalist is a native speaking scientific jargon; someone who looks directly to scientific theories to determine what exists and if the alleged scientific knowledge was obtained by reliable methods. Naturalism maintains a firm distinction between the scientist *qua* scientist and the scientist *qua* speculator: the scientist's speculations against the existence of atoms (before and after Perrins experiments) are psychologically relevant but scientifically inconsequential; primarily, the analysis of and changes in the ideological opinions of scientists over time are a matter for the history of science. (Castro 2013, p. 46)

Castro assumes a *prima facie* ironic position in this passage (the context makes it clear that Castro intends this general perspective to apply on the atoms debate in particular). Here is a philosopher of science who, based on his philosophy of science, which is squarely based on emphatic appeals to philosophy to step back and give science its way, criticizes the foremost physicists of their day, a generation which brought about scientific progress to a degree that was unthinkable before. He does so based on the conviction that he has understood the essence of science much better than these physicists and is therefore able to distinguish and cleanse pure science from its ideological and philosophical stains.[11]

[11] Furthermore, Castro (2013, pp. 45f.) calls Poincaré's conversion to atomism "philosophically bizarre" because it is either inconsistent with his philosophy of science or he had to abandon it. I hope to have shown that Castro is wrong to think that Poincaré faced this choice: his philosophy of

Of course, the fact that Castro's position is *prima facie* ironic does not constitute a move in the game of giving and asking for reasons; it does not refute Castro's general strategy. It should, however, raise eyebrows. And, I submit, the eyebrows will stay raised as one considers how Castro (2013, p. 47) distinguishes between pure science and unnecessary and obstructive philosophical preconceptions in the specific case of the atoms debate:

> From a purely scientific perspective, both histories show that, by 1900, there was scientific evidence proving the existence of atoms; the central hypothesis of kinetic theory was empirically confirmed, and it was not a useful fiction. The kinetic theory agreed with the demands of Quine's theory of confirmation: it was a theory endowed with the five theoretical benefits, in the sense that the central hypothesis was empirically confirmed and could not be jettisoned from the theory.

In this passage, Castro immediately and, for all one can tell, unconsciously equates the claim that, by 1900, the kinetic theory fulfilled the standards set by Quine's theory of confirmation with the claim that pure science had conclusive evidence for the existence of atoms by 1900.

Now, it is not in principle objectionable to be a Quinean holist with regard to theory confirmation. But it is misguided to first reproach scientists like Poincaré for bringing their own philosophies to the scientific playground, only to then bring one's own philosophy of science and call it pure science. This is exactly what Castro is doing here. So what has gone wrong? Castro criticizes these physicists for failing to appreciate the purely scientific evidence, which conclusively establishes the existence of atoms, because their view was blinded by philosophical preconception. But it turns out that, even according to Castro, this purely scientific evidence is conceived from a very specific philosophical outlook, namely a Quinean notion of empirical confirmation. By parity of reasoning, it seems therefore unfair to object, as Castro (2013, pp. 46–47) does, to Mach's refusal to take Perrin's result as conclusive evidence for atoms *solely because* Mach's refusal is based on his philosophical outlook.

---

science is much more versatile than this. Even more, I honestly don't see why a substantial change in one's philosophy of science should be termed bizarre, if it occurs in response to spectacular new developments in the sciences.

I therefore conclude that Castro cannot justify his claim that, by 1900, there was purely scientific evidence for the existence of atoms.

9.3 Conclusion: Empiricist Justificatory Monists Cannot Defend Scientific Realism

In this concluding section, I sum up my case, developed in this and the preceding chapter, for the claim that any defense of scientific realism transcends the self-set limits of the empiricist justificatory monist.

In Chapter 7, I have urged that scientific realism, as it is currently understood, directly contradicts Quine's inscrutability of reference and hence either justificatory monism or his holophrastic conception of empirical data. Furthermore, I have presented textual evidence to the conclusion that Quine never subscribed to scientific realism. The core insight of the preceding chapter is that the holistic arguments considered there fail as scientific arguments for scientific realism. I have suggested that this is because scientific realism is a non-empirical doctrine: the claim that the ontological posits of theories that satisfy one or another criterion of success exist in mind- and theory-independent reality cannot be justified by empirical means. Empirical science can only determine to what degree a theory is empirically successful.

In this present chapter, I have discussed Maddy's piecemeal realism. Maddy holds that the holistic arguments fail not because realism as such is beyond the reach of the empiricist justificatory monist, but rather because the holistic arguments are too general, too remote from actual scientific practice. This analysis leaves open the possibility of piecemeal arguments for realism regarding specific kinds of things. I have considered Maddy's main case in point, Perrin's experiments with granules in liquids at the beginning of the twentieth century. According to Maddy, the experiments directly imply realism regarding atoms. Against this, I have considered the experiments in question in some detail and argued that they do *not* directly imply realism regarding atoms; only by presupposing what I have called non-empirical bridge principles is it possible to infer realism about atoms from Perrin's experiments. Furthermore, I have shown that two key participants to the debate, Poincaré and Perrin, who both accepted versions of realism about atoms in light of Perrin's experiments, both proposed and defended elaborate versions of such bridge principles.[12]

More generally speaking, the results of the present and the preceding chapter support the claim that scientific realism is inaccessible to empiricist justificatory monists. The analysis of Psillos' NMA and of Perrin's case for the reality of the atom brought to light a justificatory gap between empirical evidence and scientific realism regarding specific kinds of entities. It is one thing to establish that a theory successfully predicts novel phenomena, or that it alone is able to causally explain a given phenomenon, but quite another thing to infer scientific realism from this. To bridge this justificatory gap, bridge principles are needed, principles that cannot be justified empirically. Whether Perrin's experiments justify realism regarding atoms, as Perrin thought, or simply provide empirical grounding, in van Fraassen's sense, for the molecular kinematic theory, is not a matter that can be decided by Perrin's experiments.

This means that the results of the present and the preceding chapter support Quine's position that empiricist justificatory monists cannot be scientific realists.

[12]However, I wish to emphasize that, while my analysis differs from Maddy's regarding the means with which the atoms debate was conducted, I entirely agree with her that the kind of realism championed by Perrin and his fellow scientists is clearly superior to the kind of abstract, holistic approach proposed by Psillos and Sankey. This can probably be explained with the interdisciplinary expertise of the scientists involved: several participants to the debate, including Perrin and Poincaré, were not only among the foremost scientists of their time, but also highly gifted philosophers. This allowed them to combine experimental findings with metaphysical theorizing in such a way as to make very compelling cases for their views. To appreciate the contrast to the contemporary scene, just imagine that Psillos would start experimenting in the CERN Hadron Collider to corroborate his realism, while leading physicists would start defending their realism in peer-reviewed philosophical journals.

References

Achinstein, Peter. 2002. Is There a Valid Experimental Argument for Scientific Realism? *The Journal of Philosophy* 99 (9): 470–495.

Burge, Tyler. 2013. *Cognition Through Understanding: Self-knowledge, Interlocution, Reasoning, Reflection.* Oxford: Oxford University Press.

Castro, Eduardo. 2013. Defending the Indispensability Argument: Atoms, Infinity and the Continuum. *Journal for General Philosophy of Science/Zeitschrift für allgemeine Wissenschaftstheorie* 44: 41–61.

Chalmers, Alan. 2011. Drawing Philosophical Lessons from Perrin's Experiments on Brownian Motion: A Response to van Fraassen. *British Journal for the Philosophy of Science* 62: 711–732.

Egg, Matthias. 2014. *Scientific Realism in Particle Physics.* Berlin: De Gruyter.

Einstein, Albert. 1905. On the Movement of Small Particles Suspended in Stationary Liquids Required by the Molecular-Kinetic Theory of Heat. *Annalen der Physik* 17: 549–560.

Gardner, Michael R. 1979. Realism and Instrumentalism in 19th-Century Atomism. *Philosophy of Science* 46 (1): 1–34.

Ivanova, Milena. 2013. Did Perrin's Experiments Convert Poincaré to Scientific Realism? *HoPos: The Journal of the International Society for the History of Philosophy of Science* 3: 1–19.

———. 2015. Conventionalism, Structuralism and Neo-Kantianism in Poincaré's Philosophy of Science. *Studies in History and Philosophy of Modern Physics* 52: 114–122.

Ladyman, James. 2011. Structural Realism Versus Standard Scientific Realism: The Case of Phlogiston and Dephlogisticated Air. *Synthese* 180 (2): 87–101.

Maddy, Penelope. 1997. *Naturalism in Mathematics.* Oxford: Oxford University Press.

———. 2007. *Second Philosophy. A Naturalistic Method.* Oxford: Oxford University Press.

Maxwell, Grover. 1962. The Ontological Status of Theoretical Entities. In *Minnesota Studies in the Philosophy of Science*, Vol. III, ed. Herbert Feigl and Grover Maxwell, 21–28. Minneapolis: University of Minnesota Press.

Nye, Mary Jo. 1972. *Molecular Reality. A Perspective on the Scientific Work of Jean Perrin.* London: MacDonald.

Ostwald, Wilhelm. 1912. *Outlines of General Chemistry*, trans. W.W. Taylor. Basingstoke: Palgrave Macmillan.

Perrin, Jean. 1909. Mouvement brownien et réalité moléculaire. *Annales de chimie et de physique* 8 (18): 5–114.

———. 1912. Les rapports de la matière et de l'éther. *Journal de physique théorique et appliquée* 2 (1): 347–360.

———. 1913. *Atomes.* Paris: Librairie Félix Alcan.

Poincaré, Henri. 1905. *La valeur de la science.* Paris: Ernest Flammarion.

Psillos, Stathis. 1999. *Scientific Realism. How Science Tracks Truth.* London: Routledge.

———. 2011. Moving Molecules Above the Scientific Horizon: On Perrin's Case for Realism. *Journal for General Philosophy of Science/Zeitschrift für allgemeine Wissenschaftstheorie* 42 (2): 339–363.

van Fraassen, Bas. 2009. The Perils of Perrin, in the Hands of Philosophers. *Philosophical Studies* 143 (1): 5–24.

Wolff, Johanna. 2015. Naturalistic Quietism or Scientific Realism? *Synthese.* https://doi.org/10.1007/s11229-015-0873-3.

10
Conclusion

In the preceding chapters, I have critically assessed epistemological naturalism, which I have conceived as consisting of naturalized epistemology and justificatory monism.

In the first part, I have started by assessing Quine's naturalized epistemology, which attempts to give a scientific explanation of the ontogenesis of successful scientific theorizing—one, notably, that supports his holophrastic conception of empirical data, and hence his empiricist justificatory monism. In Chapter 2, I have detailed the doctrinal background of Quine's naturalized epistemology. His eschatological physicalism compels him to reject final causes and intentional entities and leads him to adopt behaviorism as a method in psychology. His source-empiricism, which is further specified by his holophrastic conception of evidence, implies very limited conceptions of the evidence and the innate abilities that the infant has available to learn her first language.

Based on these principles, Quine develops an elegant and austerely physicalistic-physiological account of the infant's first steps into language in *Word and Object*. In particular, intersubjectively invariant stimulus meanings, which are directly associated with holophrastic observation sentences, provide non-verbal reference points for the infant's acquisition of

R. Gubelmann, *A Science-Based Critique of Epistemological Naturalism in Quine's Tradition*, https://doi.org/10.1007/978-3-030-24524-5_10

her first language. Unfortunately, this account is undermined by the fact that human nerve endings are non-homologous: non-homology implies that stimulus meanings, which are basically patterns of stimulation of nerve endings, cannot be intersubjectively invariant.

In Chapter 3, I have shown that this fact leads Quine on a decade-long search for a solution, which Quine thinks, in *From Stimulus to Science*, to have found in an evolutionarily shaped preestablished harmony of innate perceptual similarity standards. Such similarity standards, Quine urges, ensure that any two human beings perceive the same situations as similar, even though, for any such situation, the respective two proximal stimuli are necessarily different due to non-homology. Then, I have critically assessed this solution in two ways. On the one hand, I have argued that Quine's invocations of natural selection to explain how communication is possible probably cannot be cashed out in fully-fledged scientific explanations. On the other hand, I have urged that Quine has to, and in fact does, credit the infant with the ability to understand communicative purposes in his explanation of the infant's acquisition of her first observation sentences.

In Chapter 4, I have examined Tyler Burge's claim that there is a mature, empirically well-supported and mathematically rigorous science that has a good understanding of intentional, structured representations in perception. I have argued that Burge rightly claims that the perceptual representations postulated by explanations of perceptual psychology are structured, subject to accuracy conditions, and that they represent a distal stimulus. Hence, these representations directly contradict Quine's holophrastic conception of evidence. Furthermore, the respective explanations are also in principle irreducible to mechanistic explanations and thus equally directly contradict Quine's eschatological physicalism. However, I have also urged that the experimental support that these posits enjoy should convince Quine. When combined, these internal and external problems seriously question the basics of Quine's naturalized epistemology. I have also urged that Burge's account, rather than merely questioning Quine's approach, constitutes a step into the direction of naturalized epistemology.

In Chapter 5, I have assessed whether the naturalized epistemologist can build on this first step to take another, more significant one. I have examined the strong nativist claims that are at the core of Penelope Maddy's own version of naturalized epistemology, what she calls *Second Philoso-*

phy of Logic. In stark contrast to Quine, Maddy claims that pre-linguistic infants possess concepts, refer to objects, and are able to understand some of the things that can and cannot happen to these objects. Maddy supports this claim with evidence from habituation and preferential looking experiments with infants. I have argued that this evidence is too unspecific to support her nativist claims. In particular, the evidence provided by the experiments is too unspecific to rule out other explanations for the looking time of infants, such as Tyler Burge's, which attributes intentional representations to infants, but no understanding, nor reasoning of any sort. I have therefore concluded that naturalized epistemology has not progressed further than the step furnished by Tyler Burge, and I have also suggested that there exists a tension that might prevent it permanently from taking any further steps.

In the second part of the book, I have started by examining Quine's case for his empiricist justificatory monism. I have shown that this justificatory monism is based on verification holism, which is in turn underpinned by his holophrastic conception of empirical data. I have then argued that the problems that surfaced in Quine's naturalized epistemology question this conception of empirical data and hence his entire justificatory monism. Furthermore, I have urged that the statement of Quine's explication of justification is itself beyond the scope of justification, thus specified. Since Quine does not accept another sort of justification, this means that the central norm of the game of science, as Quine conceives it, is unjustified.

In Chapter 7, I have argued that a version of this problem applies to empiricist justificatory monism in general. Different instances of empiricist justificatory monism are individuated by their explication of empirical-scientific justification. A glance at philosophy of science textbooks shows that there are about as many such explications as there are philosophers of science. For justificatory monists, this raises the question how to defend their respective explication of justification against alternative explications. I have argued that any attempt to do so by justificatory monists necessarily begs the question. Hence, justificatory monists are unable to justify their respective explications of empirical-scientific justification.

I have then suggested that, given scientific realism, nature itself can plausibly provide a tribunal that will eventually decide on the correct explication of empirical-scientific justification and hence solve the fatal dialectic deadlock in which proponents of different justificatory monisms are stuck. I have then introduced contemporary scientific realism, and I have reconstructed Quine's reasons against this doctrine, suggesting that it contradicts, eventually, his naturalism.

In Chapters 8 and 9, I have argued that scientific realism is unavailable to empiricist justificatory monists, which vindicates Quine's position on the topic—and which constitutes a substantial systematic insight irrespective of the interests of this study. In Chapter 7, I have critically assessed the tradition of so-called no-miracle arguments ('NMA') in their most advanced form, as proposed by Stathis Psillos. My verdict on this proposal has been largely negative. The first NMA, on which Psillos' Meta-NMA is abductively based, abductively infers from the empirical success of modern scientific theories the metaphysical conclusion that the ontological posits of these theories must exist in mind- and theory-independent reality. I have pointed out that this inference is unconvincing because it falls short of the standards that Psillos himself erects to answer an important objection (what I have called Laudan-style counterexamples). Furthermore, I have argued that this central inference is in no recognizable way empirical-scientific, which implies that it is unacceptable to empiricist justificatory monists.

In Chapter 9, I have discussed Maddy's piecemeal realism, as she exemplifies it with regard to the debate on the ontological status of atoms that was provoked by Perrin's seminal experiments in the early twentieth century. By considering the central experiments in some detail, by contrasting Maddy's reading of these experiments with van Fraassen's, and by reconstructing the arguments proposed by Perrin and Poincaré for the reality of the atom, I have argued that the gap between empirical research on the one hand and scientific realism on the other cannot be closed—not even on this piecemeal level.

To conclude, I suggest the implications of the results of this study for the general prospects of empiricist justificatory monism and of naturalized epistemology.

Empiricist Justificatory Monism Is Untenable. Let me focus the results of the second part of my study in the following dilemma that presents itself to empiricist justificatory monists when it comes to the basic norms of science:

Horn 1. The first horn of the dilemma confronts those empiricist justificatory monists who, like Quine, adopt a non-realist conception of science. These justificatory monists face the problem of an irreducible justificatory pluralism: proponents of specific conceptions of empirical-scientific justification cannot exclude competing explications. This is because there is no further, neutral rational ground to stand on when scientifically assessing them, which means that each of these explications must be considered unjustified.

Horn 2. The second horn of the dilemma confronts those empiricist justificatory monists who adopt scientific realism. Adopting scientific realism potentially solves the problem of justificatory pluralism. But then, the problem arises how to defend the claim that scientific theories, in addition to being empirically successful, also correspond to theory-independent reality. I have urged that justifying this in an empirical-scientific way is impossible.

The two options presented by the dilemma are exhaustive, and they both imply that empiricist justificatory monism is untenable, since not every statement whatsoever that is necessary for science can be empirically-scientifically justified (see above, Sect. 1.1.1).

I suggest that the same underlying cause is responsible for both empiricist justificatory monists' failure to defend scientific realism and for their inability to reject alternative explications of justification (which implies their inability to justify their own respective explications). Any science that is able to establish on its own that it correctly represents some part of reality must have non-empirical access to this reality. Thanks to this non-empirical access to reality, it can compare the structure of competing scientific theories with the structure of reality and see which of them, if any, match. In other words, there is then a court of appeal that can rationally decide which of a number of competing theories is correspondence true. Similarly, justificatory monists who reject scientific realism need such a non-empirical court of appeal to solve the problem of justificatory pluralism.

However, such a tribunal would fundamentally contradict Quine's rejection of first philosophy. Quine saw this need for an independent, non-empirical tribunal clearly in the case of scientific realism, which he rejected. In contrast, he failed to realize it when it came to his claim that prediction of empirical data constitutes the ultimate checkpoint for science, which mistakenly lead him to claim that this basic norm of science can be justified within science.

Maddy's implicit empiricist justificatory monism does not save her from the dilemma. Her Second Philosopher refuses to specify generally what counts as scientifically justified and what does not (in agreement with Maddy 2007, p. 1). This way, she does not need to defend a specific conception of empirical-scientific justification, which means that she can avoid the question-begging debates among proponents of different versions of empiricist justificatory monism. However, her implicit commitment to empiricist justificatory monism means that she, too, faces the problem of justificatory pluralism in her practice (see Sect. 7.1), which means that her position, too, requires a commitment to scientific realism.

To conclude, let me mention one final option for the Second Philosopher: hope, that is, a kind of confidence that cannot be rationally defended. One might suggest that the Second Philosopher, while being demonstrably unable to enter the debate about—let alone defend—scientific realism, could still *hope* that scientific realism in general is true, and that it applies in the specific case of empirical-scientific justification, conceived as a natural kind. There are no reasons that decisively tell against assuming this position, since arguments have no power when such hope is at issue. However, it would likely be unacceptable to the epistemological naturalist, since she wants to imitate the epistemic methods of science. Certainly, hope, so conceived, does not figure among these methods. This would be mysticism, a firm belief in something that one knows one cannot argue for, rather than quietism, the decision not to enter into some debates. Maddy herself nowhere suggests any sympathy for this option.

Naturalized Epistemology Faces Formidable Challenges. The outcome of my critique of empiricist justificatory monism is that it cannot be determined exclusively on empirical grounds what it takes to empirically justify a hypothesis. This has two implications for Quine's version of naturalized epistemology. First, it implies that it is partly a non-empirical question whether the relationship between theory and evidence can be examined by examining how an infant acquires her first language: it is not a purely scientific fact that the infant and the scientist are sitting in the same epistemic boat. Maybe the scientist has qualitatively different means to justify her hypotheses and theories than the infant has to acquire his first words.

The second implication, which generalizes to all version of naturalized epistemology, is that it is determined partly non-empirically, and, if the realm of the empirical coincides with the realm of science, partly non-scientifically which scientific theories are allowed in the pursuit of naturalized epistemology: the partly non-scientific answer to the question as to what scientifically justifies a hypothesis or theory determines which theories are justified and hence qualify for employment in naturalized epistemology. This implies that, when it comes to these epistemological questions, science is answerable to a non-empirical epistemic authority, what Quine (1981, p. 72) calls a "supra-scientific tribunal", or simply "first philosophy".

With regard to Maddy's Second Philosophy of Logic, some of these partly non-empirical aspects of justifying hypotheses or theories have already surfaced (see above, Sect. 5.3.1): the nativist's case for innate concepts is abductive. Researchers such as Spelke claim that innate concepts are the best, or even the only, explanation for certain behavioral patterns that are experimentally observed. Several methodological questions surround this abduction, and none of them can be decided upon in a purely empirical way. For instance, adherents of Morgan's cannon would prefer the cognitively simpler explanations delineated in Sect. 5.3.2. Whether these simpler explanations are indeed viable is mostly an empirical question; whether the principle of Morgan's canon itself is viable, however, is at most partly empirical. This means that it is only partly an empirical question whether, in the domain of psychology, for an explanatory hypothesis to be justified requires that it withstands Morgan's canon.

Is the resulting inquiry still worthy of the term 'naturalized epistemology'? I suggest it is. My results do not imply a strict distinction between epistemology and empirical science. They do not imply that empirical science has nothing whatsoever to contribute to epistemological inquiries.

What they do imply is that, in matters epistemological (which include specification of the basic norms of science), empirical science has to be supplemented with non-empirically justified principles. 'First philosophy' seems an apt term for the reflection and research that results in these principles.

Furthermore, the results of part one of the study support the conclusion that Quine's eschatological physicalism has led his naturalized epistemology into a dead end. Once this is accepted, it comes into view that there are good, successful scientific explanations of how and when intentional states, namely perceptual representations, are produced in individuals, and how they are supported by environmental clues that reaches these individuals. This is clearly a step in the naturalized epistemologist's direction. There is no definitive reason why future, non-reductive and non-physicalistic scientific theories could not in principle make more substantial progress in the explanation of the genesis of human scientific inquiry and hence of the relationship between theory and evidence. Then, Maddy's innate logic may still receive the empirical support that, as I have suggested, it currently still lacks.

However, it may very well be that the very features of these productions of intentional states—their regularity and law-likeness—that enable these good scientific explanations also imply that they are useless in the attempt at understanding the inherently spontaneous phenomenon of scientific inquiry and its evidential support. I have suggested that Burge's objective representations do not go all the way towards a naturalized epistemology, as the respective representations, although clearly intentional, are too law-like to have a serious bearing on scientific inquiry with its inherent spontaneity. However, the same problem also arises with regard to Quine's naturalized epistemology (but not with regard to Maddy's Second Philosophy of Logic): the inherent spontaneity and creativity of scientific inquiry is very difficult to square with the process of mindless conditioning that constitutes, according to Quine, the ontogenesis of language and hence of science.

Hence, as long as it is taken to be the hallmark of successful scientific theories to predict future developments of their subject matter, and as long as future developments in the natural sciences continue to be as disruptive and unpredictable as they have been so far,[1] it seems difficult to conceive of science itself as an apt subject for scientific study. It may very well be that science itself proves to be an unnatural phenomenon in its spontaneity and unpredictability.

If, finally, these difficulties can be overcome and a science of science starts to deliver powerful, empirically supported explanations of scientific inquiry, this science of science probably confronts naturalized epistemologists with a problem in ontology that is to some extent analogous to the problem of justificatory pluralism. To see why, note again the function that Quine's physicalism performs with regard to ontology: it has the final authority on what exists. Furthermore, assume that the intentional perceptual representations postulated by perceptual psychology turn out to be part of this science of science. What kind of being do you attribute these representations to? Assuming Quinean physicalism, this question is easy to answer: ultimately, humans are physical things (compare the EN passage above, Sect. 2.1, Quine 1969, pp. 82–83); therefore, you attribute these representations to a physical being.[2]

Without such a physicalism, however, the question is perhaps not answerable without taking recourse to non-empirical ways of justification. Certainly, the respective animal, say, a honeybee, is a physical being insofar as it is made of physical stuff. However, conceived as a biological being, its identity conditions differ from its physical constitution (it can remain the same bee in spite of having replaced the entire physical stuff of which it is made). Conceived as a perceiving being, in turn, the honeybee achieves representations that are non-biological entities insofar as they are governed by a strictly non-biological teleology.

How do these various ways of conceiving the same being hang together? What justifies talk of the *same* honeybee that is conceived physically, biologically, and psychologically? Given this pluralism of scientific perspectives, and given that no single one of them can, in the manner of Quine's physicalism, claim to dominate the others in explanatory and ontological terms, it seems that a non-empirical discipline is needed to explain

[1] For a discussion of this issue, see Stanford (2015).

[2] That, as a matter of fact, physical entities cannot be credited with such representations has been shown above. The present claim is merely that physicalism is in principle able to decide such questions.

how these perspectives hang together, and how they are all perspectives on the same thing. This view of the role of philosophy has been forcefully defended by E. J. Lowe (compare, in particular, Lowe 1998), who aptly calls it 'metaphysics'. And, metaphysics, so conceived, seems to be a kind of first philosophy.

In conclusion, it seems that, in all likelihood, first philosophy in the sense of non-empirical epistemological research and reflection on scientific justification as well as in the sense of metaphysics is here to stay. The only way to eliminate it is by stipulation: to call 'science' what was formerly called 'philosophy'. But this would be naturalism only by name, unworthy of such eminent philosophers as W. V. Quine or Penelope Maddy.

References

Lowe, E.J. 1998. *The Possibility of Metaphysics: Substance, Identity, and Time.* Oxford: Oxford University Press.

Maddy, Penelope. 2007. *Second Philosophy. A Naturalistic Method.* Oxford: Oxford University Press.

Quine, Willard Van Orman. 1969. Epistemology Naturalized. In *Ontological Relativity and Other Essays*, 69–90. New York: Columbia University Press.

———. 1981. Five Milestones of Empiricism. In *Theories and Things*, 67–72. Cambridge: Harvard University Press.

Stanford, P. Kyle. 2015. Catastrophism, Uniformitarianism, and a Scientific Realism Debate That Makes a Difference. *Philosophy of Science* 82 (5): 867–878.

Index

R. Gubelmann, *A Science-Based Critique of Epistemological Naturalism in Quine's Tradition*, https://doi.org/10.1007/978-3-030-24524-5

FSC
www.fsc.org
MIX
Papier | Fördert gute Waldnutzung
FSC® C083411

Kolibri 360 GmbH
Ferdinand-Jühlke-Straße 7
99095 Erfurt, Deutschland
produktsicherheit@kolibri360.de